Collins

The Shang Maths Project

For the English National Curriculum

Practice Book 1A

Series Editor: Professor Lianghuo Fan

UK Curriculum Consultant: Paul Broadbent

William Collins' dream of knowledge for all began with the publication of his first book in 1819.

A self-educated mill worker, he not only enriched millions of lives, but also founded a flourishing publishing house. Today, staying true to this spirit, Collins books are packed with inspiration, innovation and practical expertise. They place you at the centre of a world of possibility and give you exactly what you need to explore it.

Collins. Freedom to teach.

Published by Collins, an imprint of HarperCollins*Publishers*
The News Building
1 London Bridge Street
London
SE1 9GF

Browse the complete Collins catalogue at
www.collins.co.uk

10 9 8 7 6 5 4 3 2 1

978-0-00-822607-7

Translated by Professor Lianghuo Fan, Adapted by Professor Lianghuo Fan.

British Library Cataloguing in Publication Data

A catalogue record for this publication is available from the British Library.

Series Editor: Professor Lianghuo Fan
UK Curriculum Consultant: Paul Broadbent
Publishing Manager: Fiona McGlade
In-house Editor: Nina Smith
In-house Editorial Assistant: August Stevens
Project Manager: Emily Hooton
Copy Editor: Catherine Dakin
Proofreader: Karen Williams
Cover design: Kevin Robbins and East China Normal University Press Ltd
Cover artwork: Daniela Geremia
Internal design: 2Hoots Publishing Services Ltd
Typesetting: 2Hoots Publishing Services Ltd
Illustrations: QBS
Production: Rachel Weaver
Printed and bound by CPI Group (UK) Ltd, Croydon, CR0 4YY

The Shanghai Maths Project (for the English National Curriculum) is a collaborative effort between HarperCollins, East China Normal University Press Ltd. and Professor Lianghuo Fan and his team. Based on the latest edition of the award-winning series of learning resource books, *One Lesson, One Exercise*, by East China Normal University Press Ltd. in Chinese, the series of Practice Books is published by HarperCollins after adaptation following the English National Curriculum.

Practice Book Year 1A is translated and developed by Professor Lianghuo Fan with assistance of Ellen Chen, Ming Ni, Huiping Xu and Dr. Lionel Pereira-Mendoza, with Paul Broadbent as UK Curriculum Consultant.

Contents

Chapter 1 Numbers up to 10

Chapter 2 Addition and subtraction within 10

Chapter 3 Numbers up to 20 and their addition and subtraction

1.1 Let's begin

  Learning objective Count and match objects to 10

Basic questions

1 Draw a line to match each pair.

2 Count and then match.

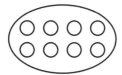　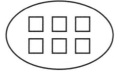　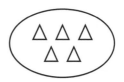　

3 Count the objects in the picture and then write the correct numbers in the boxes.

Challenge and extension question

4 How will you colour the last three faces to continue the pattern?

1.2 Let's sort (1)

 Learning objective Sort objects by two or more criteria

 Basic questions

1 Draw a line to sort each object into the correct box.

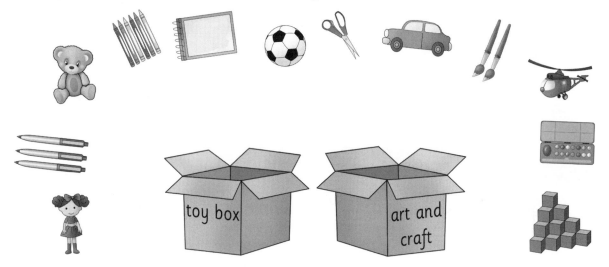

2 Draw a line to match each animal or plant to where you might find it.

3 Draw a circle around the odd one out in each set.

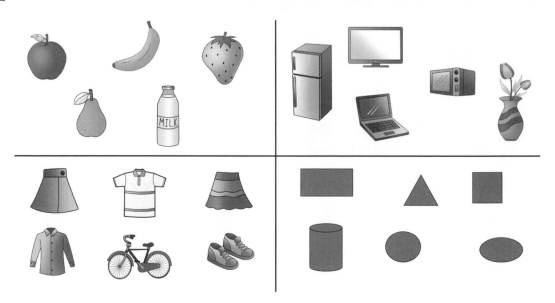

Challenge and extension question

4 Sort the toys by writing their numbers in the boxes.

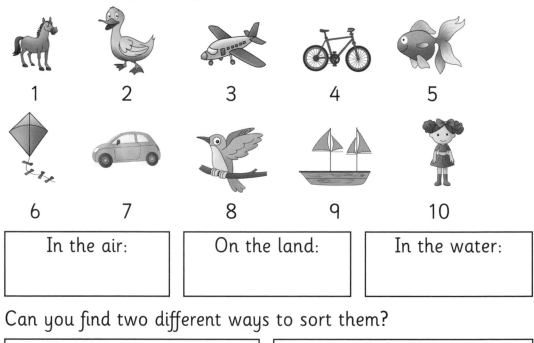

In the air:	On the land:	In the water:

Can you find two different ways to sort them?

4

1.3 Let's sort (2)

 Learning objective Sort objects in different ways

 Basic questions

1 Draw circles to sort each set into two smaller sets.

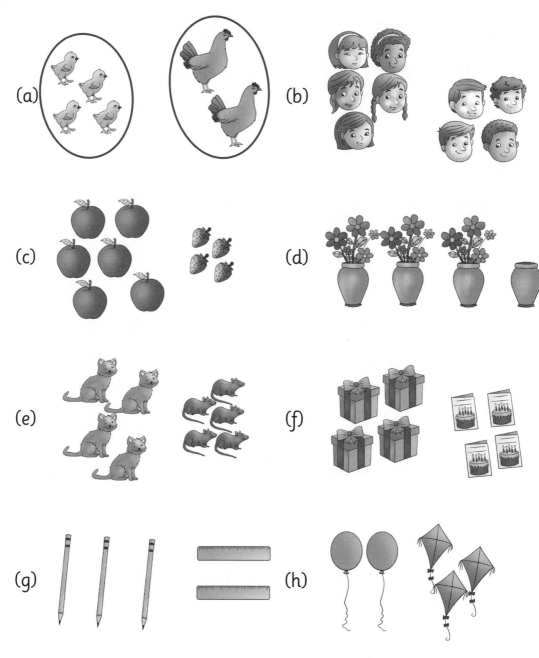

(a)

(b)

(c)

(d)

(e)

(f)

(g)

(h)

2 Find a different way to sort each set of objects.

(a)

(b)

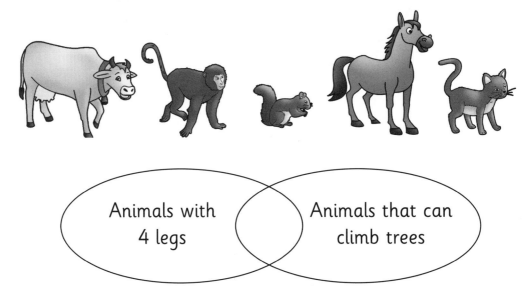

Challenge and extension question

3 Draw a line to match each animal to the correct set.

Animals with 4 legs

Animals that can climb trees

Do any of the animals belong in both sets?
Write the animals here.

1.4 Let's count (1)

 Learning objective Count and recognise numerals to 10

 Basic questions

1 Count and match each set.

one	1	two	2	three	3	four	4	five	5

six	6	seven	7	eight	8	nine	9	ten	10

Challenge and extension question

2 Count and colour. One has been done for you.

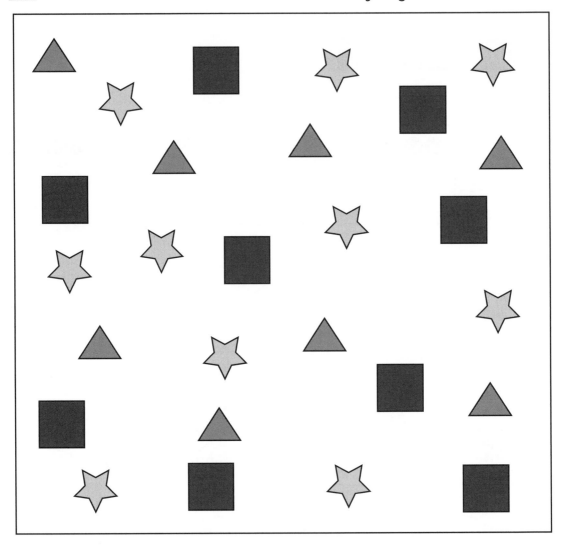

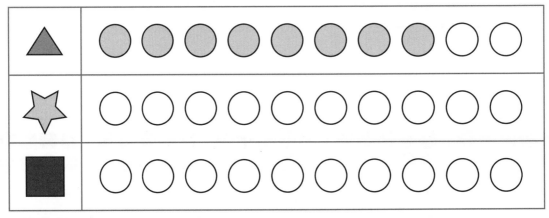

1.5 Let's count (2)

Learning objective Count and write numerals to 10

Basic questions

1 Count each set and then draw a line to match it to the correct number and number name.

1	2	3	4	5
one	two	three	four	five

Colour the correct number of dots to show the same number of balls. Write the correct number name and numeral next to each set.

Balls	Dots	Name	Numeral
⚫⚫	●●○○○	two	2
⚽⚽⚽	○○○○○		
🏈	○○○○○		
⚾⚾⚾⚾⚾	○○○○○		
🎾🎾🎾🎾	○○○○○		

Count each set and then draw a line to match it to the correct number and number name.

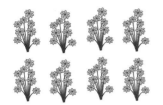  

6	7	8	9	10
six	seven	eight	nine	ten

Colour the correct number of dots to show the same number of flowers. Write the correct number name and numeral next to each set.

		eight	8

Challenge and extension question

2 Count the shapes and write the numbers in the spaces.

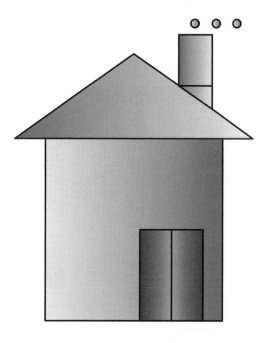

△	triangle	
○	circle	
□	square	
▭	rectangle	

1.6 Let's count (3)

 Learning objective Sort objects in different ways.

 Basic questions

1 Sort the shapes. Write their numbers in the ovals.

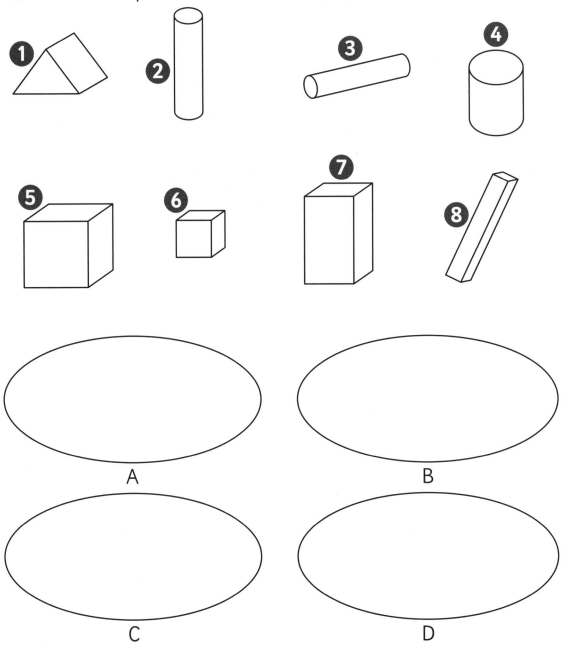

2 Count the shapes. Write the correct number of each shape in the spaces.

(a)

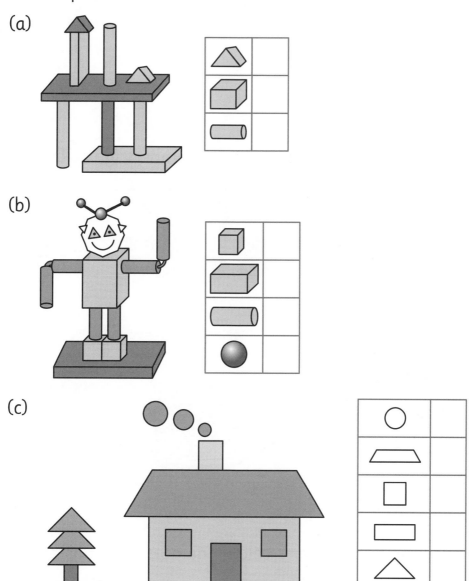

(b)

(c)

Challenge and extension question

3 Colour the cube towers that are made from 5 cubes only.

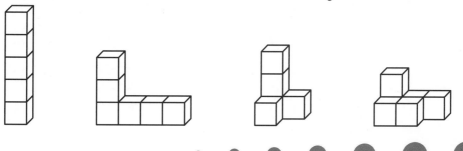

1.7 Let's count (4)

Learning objective Count groups of up to 10 objects

Basic questions

1 Count the objects and write the numbers in the boxes.

☐

☐

☐

☐

☐

☐

2 Read the numbers and colour the dots. One has been done for you.

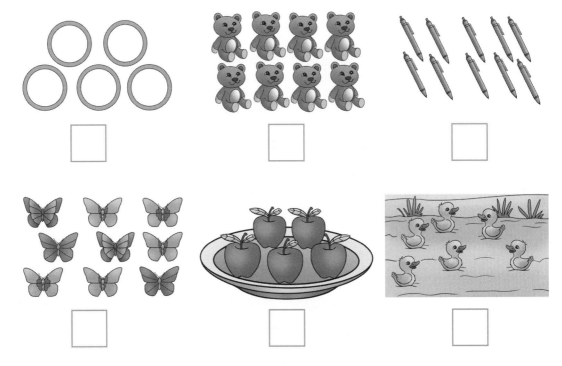

3 How many is 2 fives?

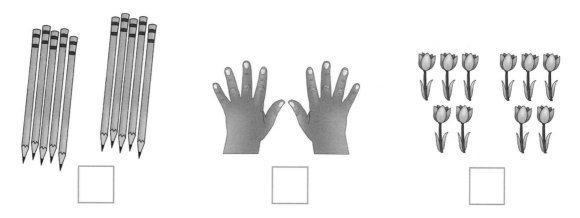

4 Count and circle 10 items in each group.

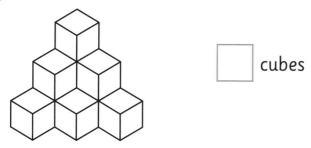

Challenge and extension question

5 Count the cubes. Write the number in the box.

◻ cubes

1.8 Let's count (5)

Learning objective Count and recognise numbers to 10, including zero

Basic questions

1 Look at the pictures and write the numbers in the boxes.

(a) How many pears are on each tree?

(b) How many apples are on each plate?

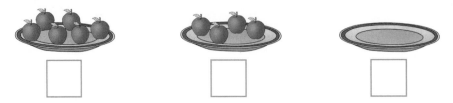

2 Count the dots and write the numbers in the boxes.

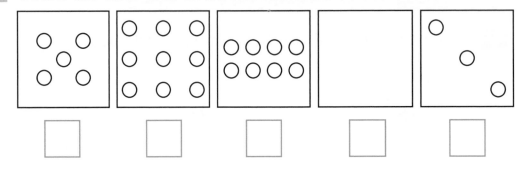

3 Draw the correct number of triangles in the spaces. One has been done for you.

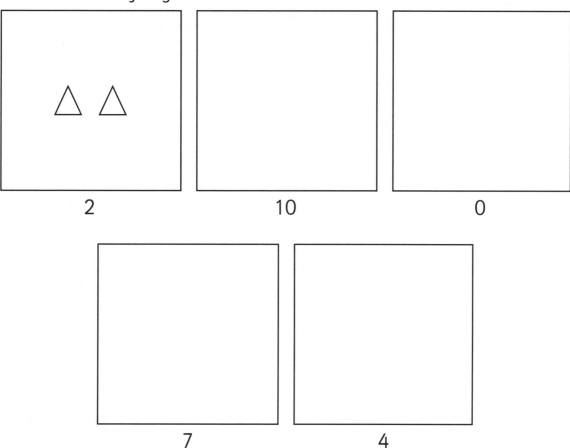

2   10   0

7   4

Challenge and extension question

4 Look at your ruler and find 0. What does it stand for?

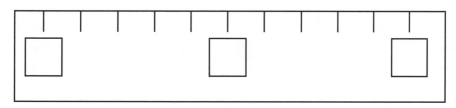

5 Write a suitable number in each box.

1.9 Let's count (6)

Learning objective Count and partition numbers to 10

Basic questions

1 Count the spots and then fill in the boxes.

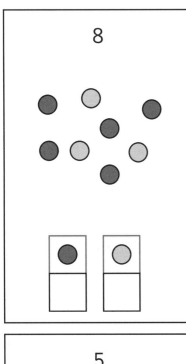

8

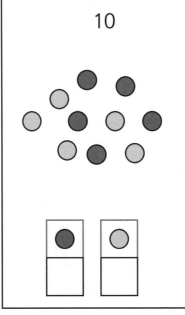

6

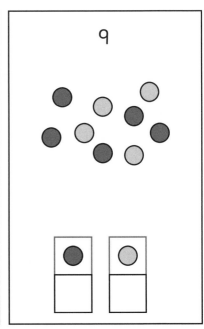

9

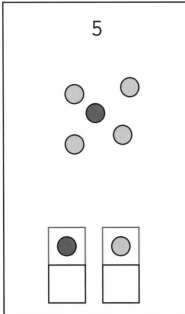

5

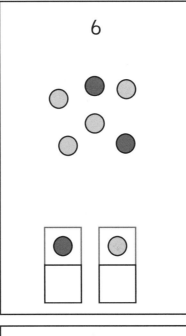

10

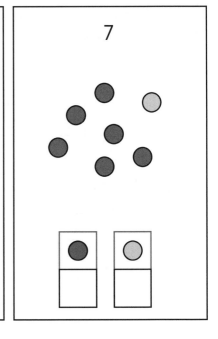

7

2 Look at the hearts in each row and then colour the dots in the same way. Write the numbers in the spaces. One has been done for you.

❤ ♡		❤	♡
♡ ♡ ♡ ♡ ♡	○ ○ ○ ○ ○		
❤ ♡ ♡ ♡ ♡	● ○ ○ ○ ○	1	4
❤ ❤ ♡ ♡ ♡	○ ○ ○ ○ ○		
❤ ❤ ❤ ♡ ♡	○ ○ ○ ○ ○		
❤ ❤ ❤ ❤ ♡	○ ○ ○ ○ ○		
❤ ❤ ❤ ❤ ❤	○ ○ ○ ○ ○		

3 Draw the eggs on the other side of each nest to show the correct total.

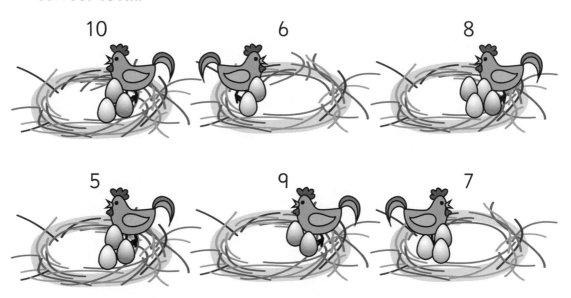

10   6   8

5   9   7

4 Show different ways to make 8 in each circle. Write the correct numbers in the boxes.

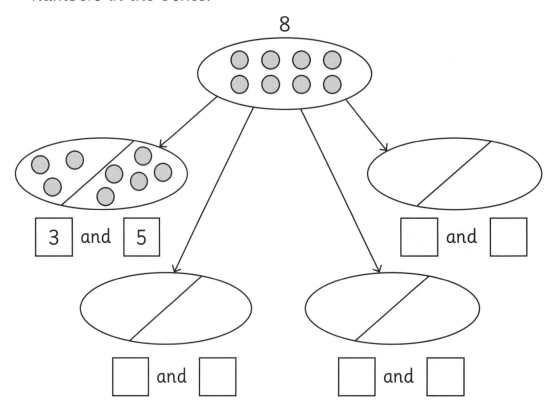

3 and 5

[ ] and [ ]

[ ] and [ ]

[ ] and [ ]

Challenge and extension question

5 Write numbers on the boxes to complete the pattern.

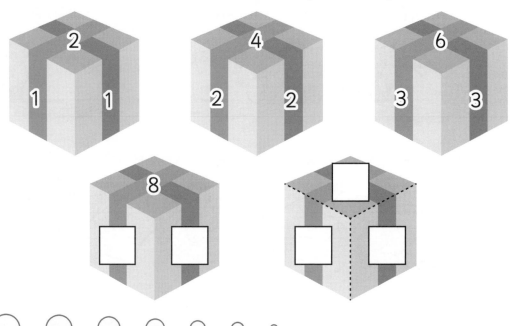

1.10 Counting and ordering numbers (1)

Learning objective Count and order numbers to 10

Basic questions

1 Look at the pictures and write your answers in the spaces.

Second
carriage

Fifth
carriage

The train has ☐ carriages.

The is in the _____ carriage

and the 🦁 is in the _____ carriage.

2 Count and then write your answers in the spaces.

(a)

left   right

There are ☐ pieces of fruit altogether.

Counting from the left, is in the _____

position, and is in the _____ position.

Counting from the right, is in the _____

position, and is the _____ position.

(b)

Counting from the left, is in the _____ position.

Counting from the right, is in the _____ position.

There are ☐ animals in total.

 is in the middle.

There are ☐ animals on its left.

There are ☐ animals on its right.

There are ☐ animals on the left of .

3 Count and colour the hearts. Start from the left.

Colour five of the hearts:

♡ ♡ ♡ ♡ ♡ ♡

Colour the fifth heart:

♡ ♡ ♡ ♡ ♡ ♡

Challenge and extension question

4 Draw the missing shapes in the boxes.

(a) Starting from left to right, ▲ is in the fourth place.

How many △ are there on its left?

▲ △ △ △ △

(b) Starting from right to left, ● is in the third place.

How many ○ are there on its right?

○ ○ ○ ●

23

1.11 Counting and ordering numbers (2)

 Learning objective Write ordinal numbers to 10

 Basic questions

1 Count the birds.

There are ☐  in total.

Count from the left and colour the fourth yellow.

Colour all the  after the sixth one yellow.

There are ☐ yellow 🐦 altogether.

2 Put the heights in order, starting with the tallest. One has been done for you.

_____ _____ __first__ _____ _____ _____

3 Write the position of each runner in the race.

1	2	3	4	5	6
place	place	place	place	place	place

4 Draw shapes according to the instructions.

Counting from the left, draw one △ in the 5th place, one ◯ in the 10th place, one ☐ in the 3rd place, and two ♡ in the 7th place.

Challenge and extension question

5 Put the pictures in the correct order. One has been done for you.

_____   _____   _____first_____

1.12 Let's compare (1)

Learning objective Compare sets to find more or fewer

Basic questions

1 Compare the two sets. Tick the box next to the set that has more.

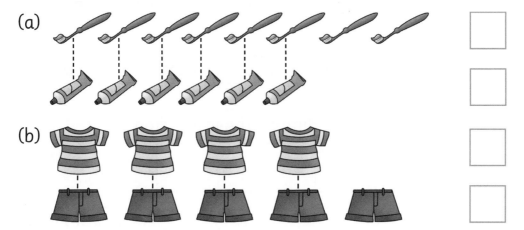

(a)

(b)

2 Compare the three sets. Tick the box next to the set that has the most.

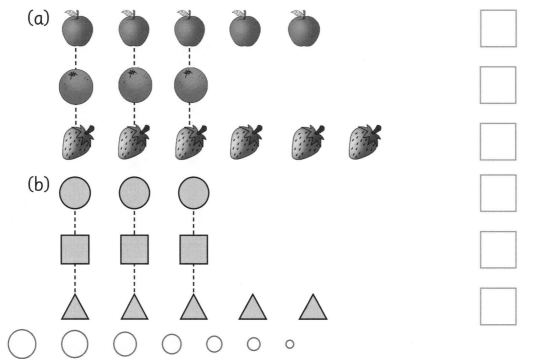

(a)

(b)

3 Compare the sets and write the missing numbers.

There are ☐ 🛢. There are ☐ 🔲.

There are ☐ fewer 🛢 than 🔲.

There are ☐ more 🔲 than 🛢.

There are ☐ ⭐. There are ☐ ★.

There are ☐ fewer ⭐ than ★.

There are ☐ more ★ than ⭐.

4 Compare and then draw.

On the first line, draw 4 ◯. _____

On the second line, draw △,
so there are 3 more than ◯. _____

On the third line, draw ☐,
so there are 2 fewer than ◯. _____

Challenge and extension question

5 Think first and then colour the longest pencil.

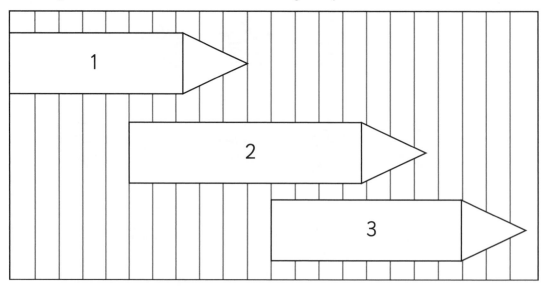

1.13 Let's compare (2)

Learning objective Compare sets using greater than or less than

Basic questions

1 Look at the pictures. Fill in the ◯ with > (greater than), < (less than) or = (equal to).

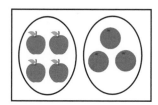  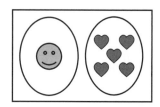

4 ◯ 3   2 ◯ 2   1 ◯ 5

2 Count the shapes and write the numbers in the boxes.
Fill in the with >, < or = .

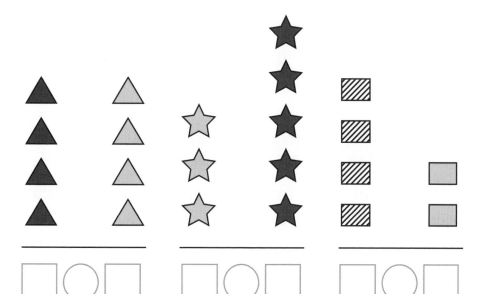

☐ ◯ ☐   ☐ ◯ ☐   ☐ ◯ ☐

3 Compare, draw and then fill in the boxes.

☐ < ☐   ☐ > ☐   ☐ = ☐

4 Fill in the ◯ with >, < or =.

4 ◯ 6   8 ◯ 5   7 ◯ 4   9 ◯ 9

8 ◯ 2   7 ◯ 8   6 ◯ 6   0 ◯ 10

5 Write a suitable number in each box.

5 < ☐   8 > ☐   7 = ☐   9 > ☐   ☐ > 7   ☐ < 10

☐ < 6   ☐ = 8   9 > ☐ > ☐   2 < ☐ < ☐

Challenge and extension question

6 Which glass of water tastes the sweetest after the cube of sugar is added? Circle the glass.

1.14 The number line

Learning objective Identify numbers to 10 on a number line

 Basic questions

1 Which of the following show a complete number line (without an ending point)? Put a ✓ in the ☐.

(a)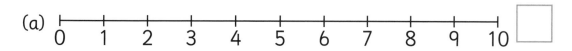

(b)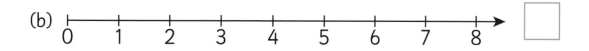

(c)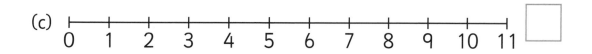

(d)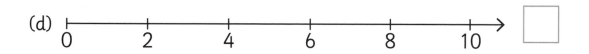

2 Write the correct numbers in the boxes on each number line.

(a)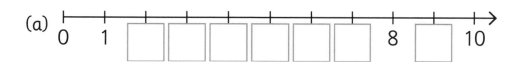

(b)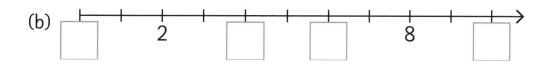

Hint: One unit on a number line is the distance between any two neighbouring numbers, for example from 0 to 1.

3 Write the correct numbers in the boxes.

(a)

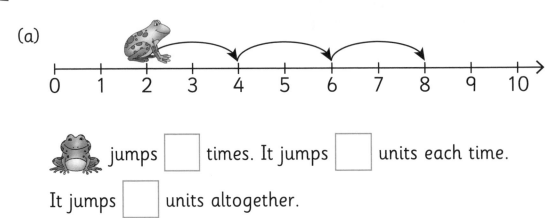

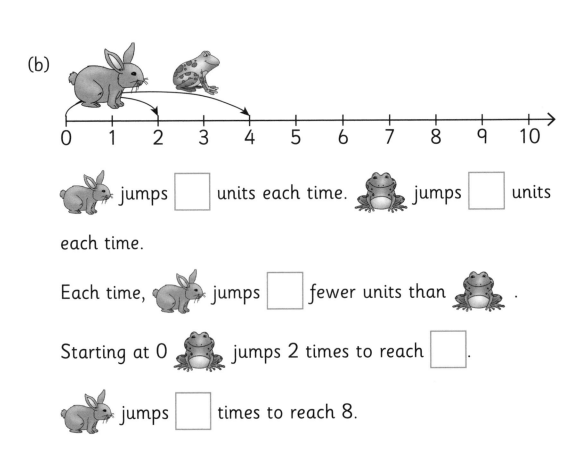

jumps ☐ times. It jumps ☐ units each time.

It jumps ☐ units altogether.

(b)

jumps ☐ units each time. jumps ☐ units each time.

Each time, jumps ☐ fewer units than .

Starting at 0 jumps 2 times to reach ☐.

jumps ☐ times to reach 8.

4 Look at this number line.

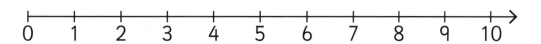

(a) Circle the numbers 7, 0, 3, 2 and 9 on the number line.

(b) Put the same five numbers in order, starting with the smallest.

(c) Put the numbers 1, 8, 5, 10 and 6 in order, starting with the largest.

Challenge and extension question

5 Write suitable numbers in the boxes. Think carefully: how many numbers are suitable to fill each box?

6 > ☐   ☐ < 8   0 < ☐ < 10   9 > ☐ > 3

Chapter 1 test

1 Draw a line to match the pairs. One has been done for you.

2 Count and draw lines to match.

6	seven	●●●●●	4	three	●●●●● ●●●●●

3 Find different ways to sort the objects into two sets.

4 Count and then write the correct answers in the spaces.

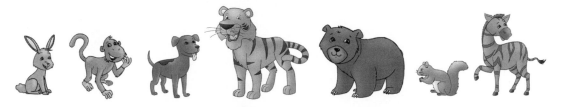

Starting from the left, is in the _____ place and

 is in the _____ place.

Starting from the right, is in the _____ place

and is in the _____ place.

There are ☐ animals altogether.

5 Count and then colour.

(a) Start from the left.

Colour four of the apples:

Colour the fourth apple:

(b) Start from the right.

Colour the fifth strawberry:

Colour five of the strawberries:

6 Put the cars in the correct order.

There are ☐ cars.

The car in first place is car number ☐.

Car number 5 is in the _____ place.

Car number 1 is in the _____ place.

The car in third place is car number _____.

7 Write numbers in the boxes to put the pictures in the correct order.

8 Write the correct numbers in the boxes to complete the number line.

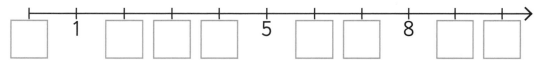

☐ 1 ☐ ☐ ☐ 5 ☐ ☐ 8 ☐ ☐

9 Write the numbers in the number line below.

(a)

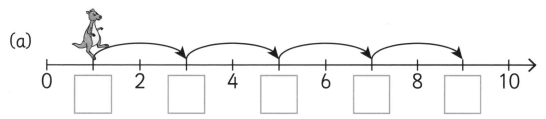

0 ☐ 2 ☐ 4 ☐ 6 ☐ 8 ☐ 10

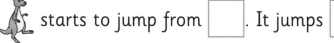 starts to jump from ☐. It jumps ☐ units each

time and finally reaches ☐.

(b)

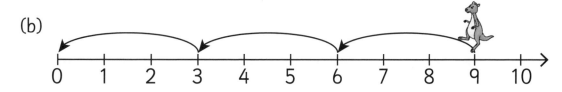

0 1 2 3 4 5 6 7 8 9 10

 jumps back from ☐. It jumps ☐ units each time

and reaches ☐.

10 Look at the pictures and compare.

There are 2 more ▲ than ☐.

There are ☐ fewer ★ than ⬤.

There are as many ☐ as ☐.

11 Write >, < or = in each ◯ and write a suitable number in each ☐.

5 ◯ 7   9 ◯ 6   4 ◯ 4   10 ◯ 5 ◯ 0

8 = ☐   ☐ < 8   7 > ☐   ☐ < 6 < ☐

☐ = ☐   ☐ < ☐   ☐ > ☐   8 ◯ ☐ ◯ 1

12 Put the numbers in order:

(a) from the smallest to the largest: 5, 8, 10, 4, 9 and 2.

☐ ☐ ☐ ☐ ☐ ☐

(b) from the largest to the smallest: 4, 7, 1, 6, 0 and 10.

☐ ☐ ☐ ☐ ☐ ☐

Chapter 2   Addition and subtraction within 10

2.1 Number bonds

 Learning objective Represent number bonds to 10

 Basic questions

1 Complete the tables. One has been done for you.

○○○○	0	4
●○○○		
●●○○		
●●●○		
●●●●		

●○○○○		
●●○○○		
●●●○○		
●●●●○		
●●●●●		

2 Circle the objects and complete the number bonds.

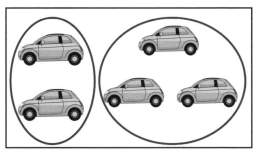

3 Fill in the boxes.

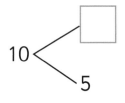 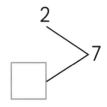 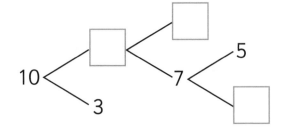

Challenge and extension question

4 Think carefully and then fill in the boxes with suitable numbers.

(a) 5 + 1 = $\boxed{4}$ + $\boxed{2}$ = $\boxed{}$ + $\boxed{}$ = $\boxed{}$ + $\boxed{}$

= $\boxed{}$ + $\boxed{}$

(b) 9 + 1 = $\boxed{8}$ + $\boxed{2}$ = $\boxed{}$ + $\boxed{}$ = $\boxed{}$ + $\boxed{}$

= $\boxed{}$ + $\boxed{}$ = $\boxed{}$ + $\boxed{}$ = $\boxed{}$ + $\boxed{}$

= $\boxed{}$ + $\boxed{}$ = $\boxed{}$ + $\boxed{}$ = $\boxed{}$ + $\boxed{}$

2.2 Addition (1)

Learning objective Add pairs of numbers with a total to 10

Basic questions

1 Fill in the boxes.

5   3   1   8   2   3   8   2   4   6

□   □   □   □   □

2 Look at the pictures and then fill in the boxes. One has been done for you.

3	+	4	=	7
4	+	3	=	7

□ + □ = □
□ + □ = □

□ + □ = □
□ + □ = □

  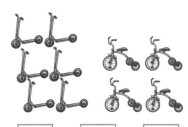

□ + □ = □
□ + □ = □

□ + □ = □
□ + □ = □

□ + □ = □
□ + □ = □

 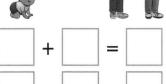 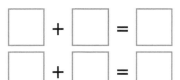

3 Complete the table.

addend	8	2	5	1	3
addend	2	4	3	9	3
sum					

addend	7	5	4	2	6
addend	1	5	3	1	4
sum					

4 Draw a line to link the additions with the same sum.

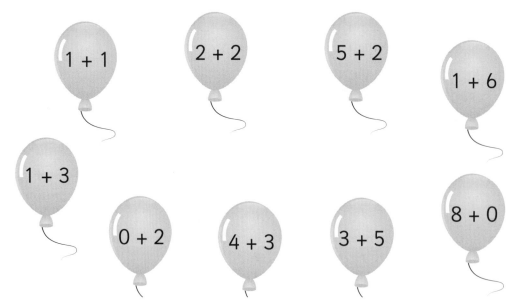

1 + 1   2 + 2   5 + 2   1 + 6

1 + 3

0 + 2   4 + 3   3 + 5   8 + 0

Challenge and extension question

5 Write addition calculations with a total of 5.

☐ + ☐ = 5   ☐ + ☐ = 5   ☐ + ☐ = 5

☐ + ☐ = 5   ☐ + ☐ = 5   ☐ + ☐ = 5

2.3 Addition (2)

Learning objective Add pairs of numbers with a total to 10

Basic questions

1 Let's think.

(a) Write each addition calculation.

○○○○
○○○○
0 + 8 = ☐

●○○○
○○○○
1 + ☐ = ☐

●●○○
○○○○
☐ + ☐ = ☐

●●●○
○○○○
☐ + ☐ = ☐

●●●●
○○○○
☐ + ☐ = ☐

●●●●
●○○○
☐ + ☐ = ☐

●●●●
●●○○
☐ + ☐ = ☐

●●●●
●●●○
☐ + ☐ = ☐

●●●●
●●●●
☐ + ☐ = ☐

(b) Colour the dots to show different sums of 9.

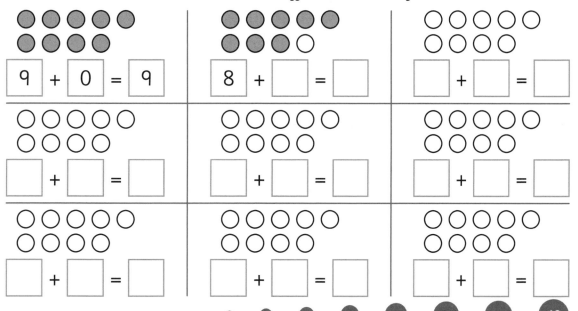

●●●●●
●●●●
9 + 0 = 9

●●●●●
●●●○
8 + ☐ = ☐

○○○○○
○○○○
☐ + ☐ = ☐

○○○○○
○○○○
☐ + ☐ = ☐

○○○○○
○○○○
☐ + ☐ = ☐

○○○○○
○○○○
☐ + ☐ = ☐

○○○○○
○○○○
☐ + ☐ = ☐

○○○○○
○○○○
☐ + ☐ = ☐

○○○○○
○○○○
☐ + ☐ = ☐

2 Write the numbers in the boxes to complete each addition calculation.

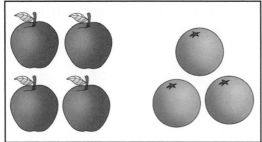 

4 + 3 = ☐   ☐ + ☐ = ☐

 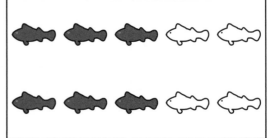

☐ + ☐ = ☐   ☐ + ☐ = ☐

3 Work out the calculations.

3 + 5 = ☐   4 + 6 = ☐   2 + 5 = ☐

0 + 4 = ☐   4 + 4 = ☐   5 + 1 = ☐

2 + 2 = ☐   6 + 2 = ☐   2 + 8 = ☐

3 + 3 = ☐   6 + 3 = ☐   5 + 4 = ☐

4 Use the picture to write different addition sentences.

☐ + ☐ = ☐ _____

☐ + ☐ = ☐ _____

☐ + ☐ = ☐ _____

☐ + ☐ = ☐ _____

2.4 Addition (3)

 Learning objective Add to 10 by combining

 Basic questions

1 Look at the pictures and write the addition sentences.

There are 6 rabbits.   Another 4 rabbits join them.   How many rabbits are there now?

□ ○ □ = □

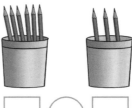

□ ○ □ = □

□ ○ □ = □

□ ○ □ = □

□ ○ □ = □

2 Write the missing number or symbol in each box. One has been done for you.

$2 \xrightarrow{+5} \boxed{7}$ $1 \xrightarrow{+2} \boxed{}$ $4 \xrightarrow{+4} \boxed{}$

$5 \xrightarrow{+1} \boxed{}$ $6 \xrightarrow{+3} \boxed{}$ $7 \xrightarrow{+0} \boxed{}$

$\boxed{} \xrightarrow{+5} 10$ $\boxed{} \xrightarrow{+7} 8$ $\boxed{} \xrightarrow{+6} 9$

$7 \xrightarrow{+\boxed{}} 10$ $2 \xrightarrow{+\boxed{}} 8$ $5 \xrightarrow{+\boxed{}} 5$

3 Write the answer in each box.

$6 + 2 = \boxed{}$ $3 + 5 = \boxed{}$ $4 + 4 = \boxed{}$

$5 + 0 = \boxed{}$ $7 + 1 = \boxed{}$ $5 + 4 = \boxed{}$

$2 + 6 = \boxed{}$ $3 + 2 = \boxed{}$ $0 + 10 = \boxed{}$

$7 + 3 = \boxed{}$ $3 + 3 = \boxed{}$ $0 + 4 = \boxed{}$

$1 + 9 = \boxed{}$ $6 + 3 = \boxed{}$

$0 + 8 = \boxed{}$ $2 + 4 = \boxed{}$

Challenge and extension question

4 Think carefully and then write the correct number in each box.

(a) If ⭐ + 🌼 = 8 and 5 + 🌼 = 7, then

⭐ = ☐ and 🌼 = ☐

(b) If ◎ = ⭐ ⭐ ⭐ and ⭐ = △ △ , then

◉ = ☐

2.5 Let's talk and calculate (I)

 Learning objective Write addition sentences with totals to 10

 Basic questions

1 Look at the pictures and write the addition sentences.

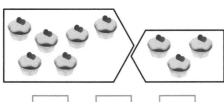

☐ + ☐ = ☐   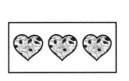

☐ + ☐ = ☐

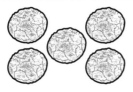

☐ + ☐ = ☐   

☐ + ☐ = ☐

---

3  left the car park, | then 6 more followed them.

How many cars left the car park?

☐ + ☐ = ☐

4  were eaten,   2 are left.

How many were there at first?

☐ + ☐ = ☐

Sort by colour:

☐ + ☐ = ☐

☐ + ☐ = ☐

Sort by size:

☐ + ☐ = ☐

☐ + ☐ = ☐

2 Write the missing number in each box.

$1 \xrightarrow{+6} \boxed{7}$ $2 \xrightarrow{+4} \boxed{}$ $3 \xrightarrow{+5} \boxed{}$

$4 \xrightarrow{+4} \boxed{}$ $6 \xrightarrow{+3} \boxed{}$ $2 \xrightarrow{+8} \boxed{}$

$7 \xrightarrow{+3} \boxed{}$ $3 \xrightarrow{+4} \boxed{}$ $6 \xrightarrow{+4} \boxed{}$

$2 \xrightarrow{+\boxed{}} 9$ $0 \xrightarrow{+\boxed{}} 6$ $4 \xrightarrow{+\boxed{}} 5$

$\boxed{} \xrightarrow{+2} 5$ $\boxed{} \xrightarrow{+3} 8$ $\boxed{} \xrightarrow{+1} 10$

3 Work out the addition calculations.

$2 + 3 = \boxed{}$ $5 + 4 = \boxed{}$ $1 + 4 = \boxed{}$ $6 + 2 = \boxed{}$

$4 + 0 = \boxed{}$ $3 + 7 = \boxed{}$ $4 + 2 = \boxed{}$ $5 + 4 = \boxed{}$

$3 + 3 = \boxed{}$ $5 + 2 = \boxed{}$ $3 + 6 = \boxed{}$ $0 + 7 = \boxed{}$

$6 + 4 = \boxed{}$ $3 + 4 = \boxed{}$ $4 + 5 = \boxed{}$ $8 + 2 = \boxed{}$

4 Write the two addition number sentences for each domino.

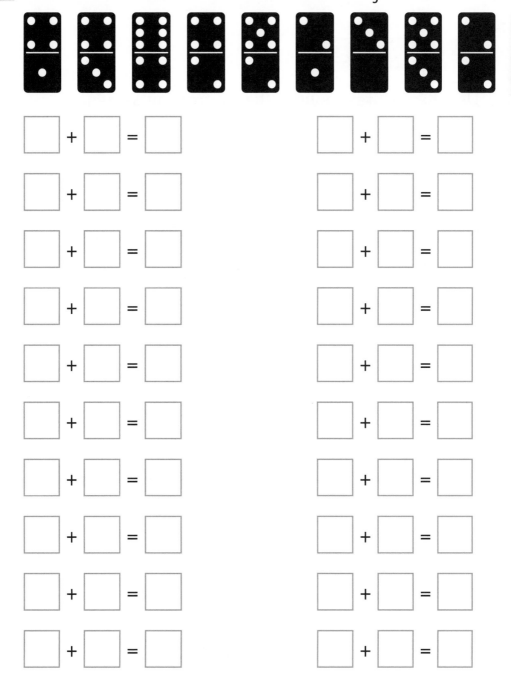

☐ + ☐ = ☐   ☐ + ☐ = ☐

☐ + ☐ = ☐   ☐ + ☐ = ☐

☐ + ☐ = ☐   ☐ + ☐ = ☐

☐ + ☐ = ☐   ☐ + ☐ = ☐

☐ + ☐ = ☐   ☐ + ☐ = ☐

☐ + ☐ = ☐   ☐ + ☐ = ☐

☐ + ☐ = ☐   ☐ + ☐ = ☐

☐ + ☐ = ☐   ☐ + ☐ = ☐

☐ + ☐ = ☐   ☐ + ☐ = ☐

☐ + ☐ = ☐   ☐ + ☐ = ☐

Which domino has only one addition number sentence?

2.6 Subtraction (1)

Learning objective Subtract within 10 by taking away

Basic questions

1 Use the pictures to write the subtraction sentences.

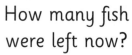

There were 6 fish on the plates.

The cat ate 2 fish.

How many fish were left now?

☐ ◯ ☐ = ☐

There are 6 birds in the tree.

4 birds fly away.

How many birds are in the tree now?

☐ ◯ ☐ = ☐

There were 7 balloons in a bunch.

2 balloons floated away.

How many balloons were left now?

□ ○ □ = □

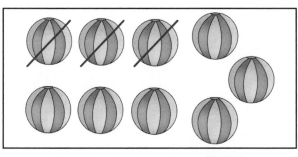

□ ○ □ = □

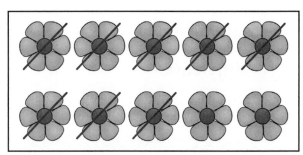

□ ○ □ = □

2 Complete the subtraction calculations.

8 – 4 = □   10 – 9 = □   6 – 4 = □   2 – 2 = □

4 – 3 = □   5 – 2 = □   9 – 4 = □   6 – 1 = □

10 – 5 = □   8 – 7 = □   7 – 5 = □   8 – 6 = □

5 – 5 = □   2 – 0 = □   4 – 1 = □   9 – 6 = □

3 Write the missing numbers in the spaces.

Remember: minuend − subtrahend = difference

minuend	8	10	5	9	3
subtrahend	2	4	3	1	3
difference					

minuend	7	5	4	2	6
subtrahend	1	5	3	1	4
difference					

Challenge and extension question

4 Write subtraction calculations with a difference of 5.

☐ − ☐ = 5   ☐ − ☐ = 5   ☐ − ☐ = 5

☐ − ☐ = 5   ☐ − ☐ = 5   ☐ − ☐ = 5

2.7 Subtraction (2)

 Learning objective Subtract within 10 using known addition bonds

 Basic questions

1 Fill in the boxes. One has been done for you.

(5) → (3) (2)   (8) → (4) [ ]   (7) → (5) [ ]   (9) → (6) [ ]   (10) → (2) [ ]

2 Look at the pictures and fill in the boxes.

There are 7 pieces of fruit altogether.

 4 apples   How many oranges?

[ ] ◯ [ ] = [ ]

There are 10 vehicles in total.

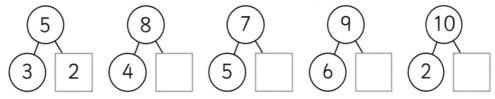 6 cars   How many vans?

[ ] ◯ [ ] = [ ]

5 footballs  How many basketballs?

There are 8 balls in total.

[ ] ◯ [ ] = [ ]

 3 lollipops   How many candy canes?

There are 10 sweets altogether.

[ ] ◯ [ ] = [ ]

How many apples are inside?

There are 9 apples altogether.

[ ] ◯ [ ] = [ ]

How many books are inside?

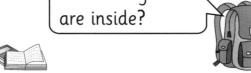 

There are 7 books in total.

[ ] ◯ [ ] = [ ]

3 Fill in the missing numbers and symbols.

5 $\xrightarrow{-2}$ ☐   4 $\xrightarrow{-3}$ ☐   8 $\xrightarrow{-6}$ ☐

9 $\xrightarrow{-1}$ ☐   6 $\xrightarrow{-5}$ ☐   10 $\xrightarrow{-2}$ ☐

6 $\xrightarrow{-3}$ ☐   7 $\xrightarrow{-7}$ ☐   9 $\xrightarrow{-4}$ ☐

10 $\xrightarrow{\boxed{-\ }}$ 6   8 $\xrightarrow{\boxed{-\ }}$ 5   7 $\xrightarrow{\boxed{-\ }}$ 1

4 Complete the subtraction calculations.

6 − 2 = ☐   4 − 4 = ☐   5 − 0 = ☐   7 − 2 = ☐

7 − 6 = ☐   9 − 4 = ☐   6 − 3 = ☐   3 − 2 = ☐

5 − 2 = ☐   7 − 3 = ☐   9 − 6 = ☐   10 − 5 = ☐

9 − 1 = ☐   6 − 5 = ☐   8 − 8 = ☐   2 − 0 = ☐

 Challenge and extension question

5 Think carefully and then fill in the boxes. Remember to look for a pattern to help you.

2 − 0 = ☐ = 3 − 1 = ☐ =

☐ − ☐ = ☐ = ☐ − ☐ = ☐ =

☐ − ☐ = ☐ = ☐ − ☐ = ☐ =

☐ − ☐ = ☐ = ☐ − ☐ = ☐

2.8 Subtraction (3)

Learning objective Subtract within 10 using known addition bonds

Basic questions

1 Look at the pictures and write the subtraction sentences.

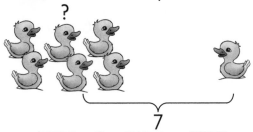

7

☐ ◯ ☐ = ☐

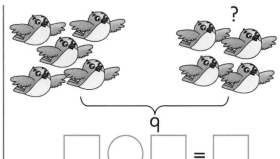

9

☐ ◯ ☐ = ☐

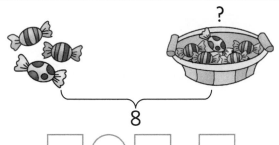

8

☐ ◯ ☐ = ☐

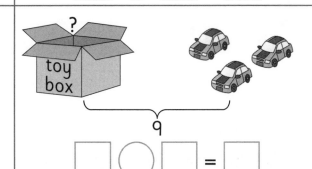

9

☐ ◯ ☐ = ☐

☐ ◯ ☐ = ☐

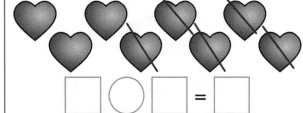

☐ ◯ ☐ = ☐

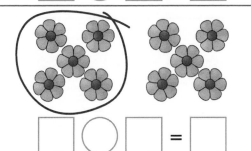

☐ ◯ ☐ = ☐

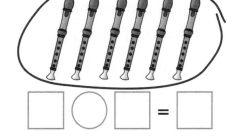

☐ ◯ ☐ = ☐

2 Write the correct number in each box.

8 $\xrightarrow{-4}$ ☐   10 $\xrightarrow{-5}$ ☐   6 $\xrightarrow{-3}$ ☐

5 $\xrightarrow{-2}$ ☐   7 $\xrightarrow{-3}$ ☐   8 $\xrightarrow{-2}$ ☐

☐ $\xrightarrow{-2}$ 5   ☐ $\xrightarrow{-3}$ 0   ☐ $\xrightarrow{-1}$ 9

3 Complete the subtraction calculations.

8 – 5 = ☐   10 – 4 = ☐   4 – 4 = ☐   6 – 0 = ☐

6 – 3 = ☐   8 – 7 = ☐   9 – 6 = ☐   8 – 3 = ☐

8 – 1 = ☐   5 – 2 = ☐   7 – 6 = ☐   10 – 7 = ☐

 Challenge and extension question

4 Look at the picture and then write the number sentences.

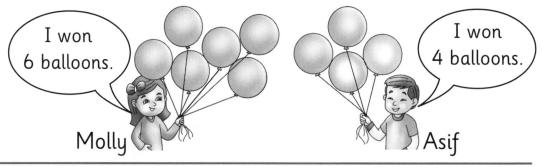

I won 6 balloons.

I won 4 balloons.

Molly   Asif

How many balloons did the children win altogether?   ☐ ◯ ☐ = ☐

How many more balloons than Asif did Molly win?   ☐ ◯ ☐ = ☐

How many more balloons does Asif need to win so that he has the same number as Molly?   ☐ ◯ ☐ = ☐

2.9 Let's talk and calculate (II)

Learning objective Interpret word problems using subtraction facts to 10

Basic questions

1 Read each number story and write the subtraction sentence to match.

There were 6 in the tree.
2 flew away.
How many are left in the tree?

☐ ◯ ☐ = ☐

There were 10 in the car park.
7 drove away.
How many  are left now?

☐ ◯ ☐ = ☐

There are 8 ❘ and ╱ altogether.

5 of them are ❘ .

How many ╱ are there?

☐ ◯ ☐ = ☐

There are 7 and in total.

There are 4 .

How many are there?

☐ ◯ ☐ = ☐

There are 9 swans in the lake.

Four of them are  .

How many  are there in the lake?

☐ ◯ ☐ = ☐

There are 7 pairs of  and  in total.

Five of them are .

How many pairs of ❤ are there?

☐ ◯ ☐ = ☐

2 Write the correct number in each box.

10 $\xrightarrow{-6}$ ☐   8 $\xrightarrow{-4}$ ☐   7 $\xrightarrow{-5}$ ☐   7 $\xrightarrow{-4}$ ☐

6 $\xrightarrow{-3}$ ☐   9 $\xrightarrow{-8}$ ☐   ☐ $\xrightarrow{-1}$ 6   5 $\xrightarrow{-\boxed{}}$ 4

8 $\xrightarrow{-\boxed{}}$ 4   7 $\xrightarrow{-\boxed{}}$ 5   ☐ $\xrightarrow{-2}$ 7   4 $\xrightarrow{-\boxed{}}$ 0

☐ $\xrightarrow{-2}$ 6   4 $\xrightarrow{-\boxed{}}$ 3   9 $\xrightarrow{-\boxed{}}$ 9

3 Complete the following calculations.

3 − 0 = ☐   5 + 4 = ☐   8 − 8 = ☐   9 − 5 = ☐

4 + 6 = ☐   9 − 3 = ☐   5 + 2 = ☐   10 − 3 = ☐

7 + 2 = ☐   10 − 7 = ☐   3 − 2 = ☐   0 + 1 = ☐

8 − 5 = ☐   5 + 5 = ☐   4 − 0 = ☐   4 − 2 = ☐

 Challenge and extension question

4 What does each shape stand for?

If △ + ☆ = 10 and 8 − ☆ = 2, then

△ = ☐ and ☆ = ☐ .

If ◆ − ● = 7 and ● + 3 = 5, then

◆ = ☐ and ● = ☐ .

2.10 Addition and subtraction

Learning objective Use the inverse relationship between addition and subtraction with numbers to 10

Basic questions

1 Write the number sentence for each picture.

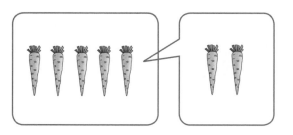

5 + ☐ = ☐

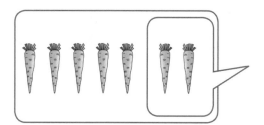

7 − ☐ = ☐

☐ + ☐ = ☐

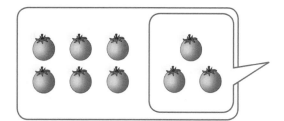

☐ − ☐ = ☐

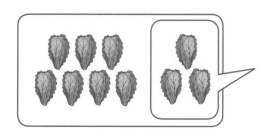

☐ − ☐ = ☐

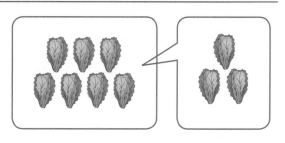

☐ + ☐ = ☐

2 Write the missing number in each box.

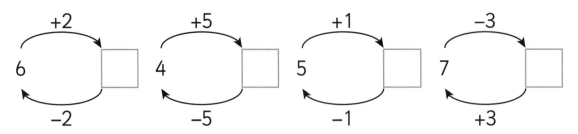

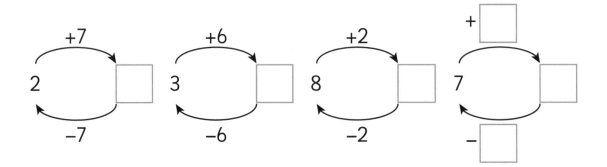

3 Fill in the boxes.

4 + 1 = ☐

5 − 4 = ☐

5 − 1 = ☐

4 + 6 = ☐

10 − 4 = ☐

10 − 6 = ☐

2 + 3 = ☐

5 − 2 = ☐

5 − 3 = ☐

8 + 1 = ☐

9 − ☐ = ☐

9 − ☐ = ☐

3 + 7 = ☐

10 − ☐ = ☐

10 − ☐ = ☐

5 + 2 = ☐

☐ − ☐ = ☐

☐ − ☐ = ☐

4 Write the correct number in each box.

$$3 \xrightarrow{+2} \boxed{} \xrightarrow{+4} \boxed{}$$

$$9 \xrightarrow{-2} \boxed{} \xrightarrow{-5} \boxed{}$$

$$4 \xrightarrow{+6} \boxed{} \xrightarrow{-5} \boxed{}$$

$$10 \xrightarrow{-7} \boxed{} \xrightarrow{+3} \boxed{}$$

$$5 \xrightarrow{+5} \boxed{} \xrightarrow{-6} \boxed{}$$

$$4 \xrightarrow{-4} \boxed{} \xrightarrow{+10} \boxed{}$$

Challenge and extension question

5 Write subtraction calculations with a difference of 3.

$$\boxed{} - \boxed{} = 3 \qquad \boxed{} - \boxed{} = 3$$

$$\boxed{} - \boxed{} = 3 \qquad \boxed{} - \boxed{} = 3$$

$$\boxed{} - \boxed{} = 3 \qquad \boxed{} - \boxed{} = 3$$

$$\boxed{} - \boxed{} = 3 \qquad \boxed{} - \boxed{} = 3$$

2.11 Addition and subtraction using a number line

Learning objective Add and subtract numbers to 10 using a number line

 Basic questions

1 Fill in the boxes to complete the number lines.

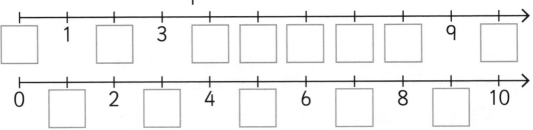

2 Calculate using a number line.

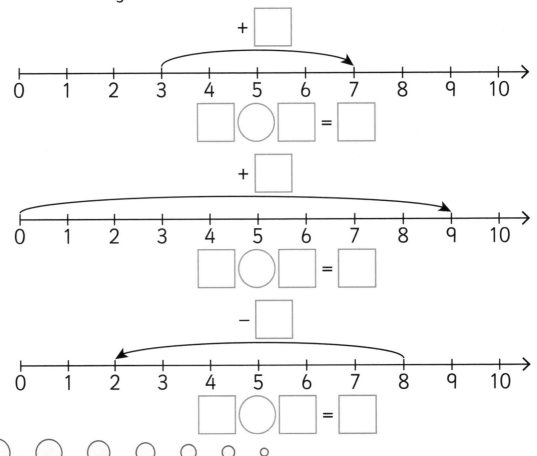

3 Complete the following calculations.

4 + 4 = ☐   9 − 3 = ☐   10 − 4 = ☐   3 + 6 = ☐

10 − 6 = ☐   5 + 5 = ☐   9 + 1 = ☐   9 − 7 = ☐

3 + 6 = ☐   3 + 7 = ☐   4 + 6 = ☐   3 + 4 = ☐

10 − 5 = ☐   7 − 7 = ☐   10 − 8 = ☐   8 − 4 = ☐

4 Write the missing numbers.

☐ + 4 = 8   2 + ☐ = 10   7 − ☐ = 3

3 + ☐ = 6   ☐ + 8 = 8   ☐ − 6 = 3

5 + ☐ = 9   ☐ − 3 = 6   8 − ☐ = 10 − 4

Challenge and extension question

5 Look at each number sentence. Draw it on the number line and find the answer.

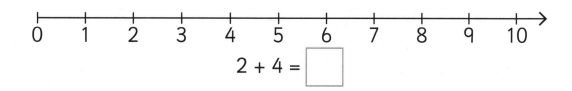

2 + 4 = ☐

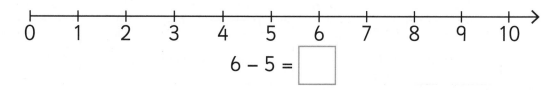

6 − 5 = ☐

2.12 Games of number 10

 Learning objective Solve problems for pairs of numbers that total 10

 Basic questions

1 Draw lines to join pairs of numbers that make 10.

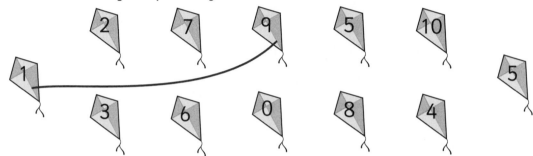

2 Write suitable numbers into the boxes.

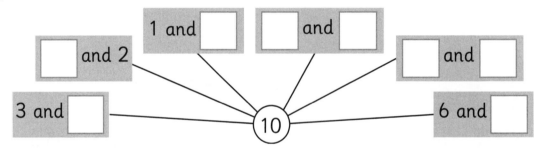

3 Which key is for which lock? Draw lines to link them.

4 Draw lines to match. One has been done for you.

2 + 8	2 + 7
4 + 5	5 + 3
2 + 6	3 + 4
9 – 2	3 + 7

7 – 5	7 – 4
9 – 3	6 – 4
8 – 4	9 – 5
6 – 3	8 – 2

5 Fill in the boxes with suitable numbers.

$5 + \boxed{} = 10$ $10 - \boxed{} = 0$ $9 + \boxed{} = 10$ $10 - \boxed{} = 1$

$3 + \boxed{} = 10$ $\boxed{} + 2 = 10$ $10 = 7 + \boxed{}$ $\boxed{} - 5 = 5$

$\boxed{} + 8 = 10$ $3 + \boxed{} = 7$ $4 + \boxed{} = 8$ $\boxed{} - \boxed{} = 3$

6 Write >, < or = in each $\bigcirc$.

$7 + 3 \bigcirc 8$ $5 + 4 \bigcirc 7$ $2 + 7 \bigcirc 5 + 3$

$4 - 2 \bigcirc 6$ $9 - 3 \bigcirc 6$ $5 + 3 \bigcirc 9 + 0$

$2 + 8 \bigcirc 10$ $8 - 5 \bigcirc 10$ $6 + 4 \bigcirc 5 + 5$

Challenge and extension question

7 Write addition sentences with totals of 10.

$10 = \boxed{} + \boxed{}$ $= \boxed{} + \boxed{}$

$= \boxed{} + \boxed{}$ $= \boxed{} + \boxed{}$

$= \boxed{} + \boxed{}$ $= \boxed{} + \boxed{} + \boxed{}$

2.13 Adding three numbers

Learning objective Add three numbers with totals to 10

Basic questions

1 Look at each picture and write the number sentence.

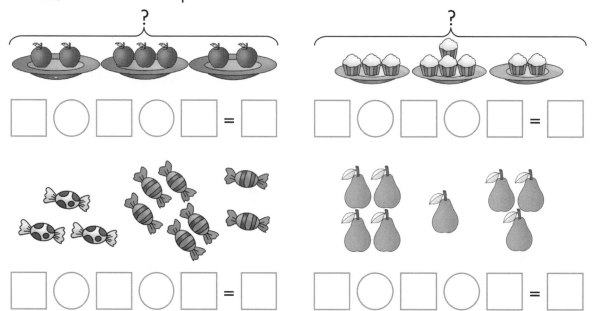

2 Look at the number lines and complete the calculations.

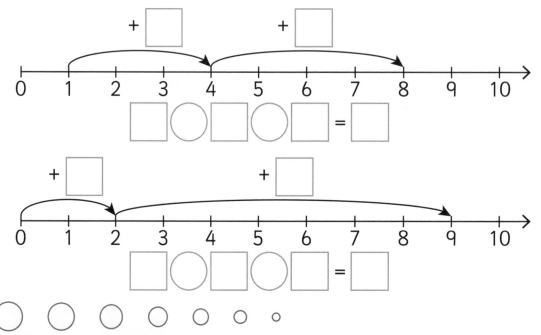

3 Complete the calculations.

1 + 3 + 3 = ☐   0 + 3 + 4 = ☐   8 + 0 + 2 = ☐

3 + 4 + 0 = ☐   2 + 2 + 1 = ☐   1 + 6 + 2 = ☐

2 + 1 + 6 = ☐   5 + 3 + 2 = ☐   2 + 7 + 1 = ☐

2 + 4 + 1 = ☐   4 + 2 + 4 = ☐   1 + 4 + 5 = ☐

4 Write the number sentences.

(a) 3 children were playing football in the playground. Another 2 children joined them. Then 5 more children joined the game. How many children played football?

Number sentence: _____

(b) Amina has 4 red pencils and 2 blue pencils. She has the same number of green pencils as blue ones. How many pencils does she have altogether?

Number sentence: _____

Challenge and extension question

5 Think carefully and then write the answers in the boxes.

2 + 3 + ☐ = 9
☐

☐ + 1 + 2 = 6
☐

4 + ☐ + 4 = 10
☐

☐ + 3 + 5 = 9
☐

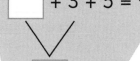

69

2.14 Subtracting three numbers

Learning objective Subtract two smaller numbers from a number to 10

Basic questions

1 Look at each picture and write the number sentence.

☐ ○ ☐ ○ ☐ = ☐   ☐ ○ ☐ ○ ☐ = ☐

2 Look at the number lines and complete the calculations.

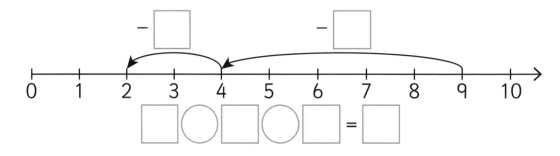

–☐   –☐

☐ ○ ☐ ○ ☐ = ☐

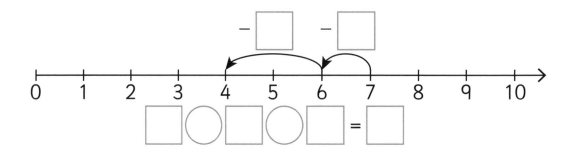

–☐ –☐

☐ ○ ☐ ○ ☐ = ☐

3 Complete the subtraction calculations.

$5 - 3 - 1 = \square$ $6 - 3 - 3 = \square$ $8 - 4 - 2 = \square$

$6 - 4 - 0 = \square$ $7 - 2 - 1 = \square$ $9 - 3 - 5 = \square$

$8 - 7 - 1 = \square$ $10 - 3 - 4 = \square$ $10 - 7 - 3 = \square$

$8 - 2 - 2 = \square$ $7 - 2 - 3 = \square$ $6 - 3 - 1 = \square$

$10 - 6 - 2 = \square$ $6 - 5 - 1 = \square$ $4 - 2 - 2 = \square$

4 Read the problems and then write the number sentences.

(a) 9 birds were in the tree. 5 birds flew away. Then 2 more flew away. How many birds are still in the tree?

Number sentence: _____

(b) There are 9 balloons. 3 of them are blue. Another 4 are red. The rest are yellow. How many balloons are yellow?

Number sentence: _____

Challenge and extension question

5 Think carefully and then write the answers in the boxes.

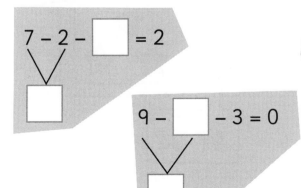

$7 - 2 - \square = 2$

$9 - \square - 3 = 0$

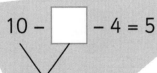

$10 - \square - 4 = 5$

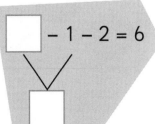

$\square - 1 - 2 = 6$

71

2.15 Mixed addition and subtraction

Learning objective Add and subtract three numbers within 10

Basic questions

1 Look at the pictures and write the number sentences.

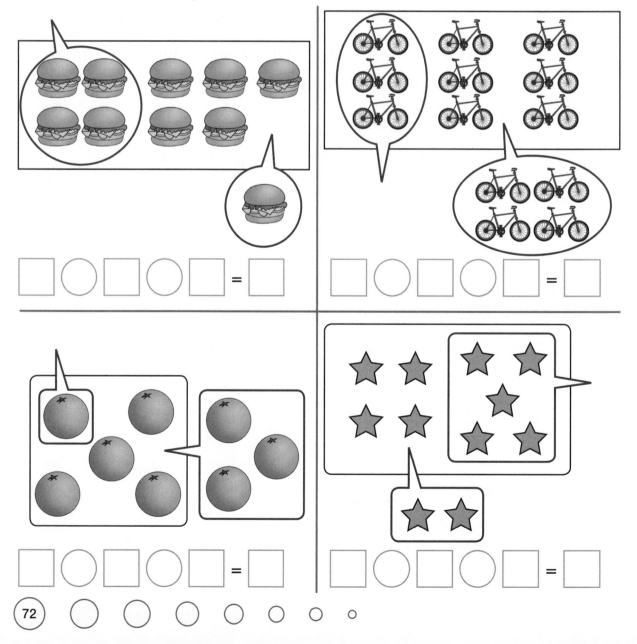

2 Look at the number lines and complete the calculations.

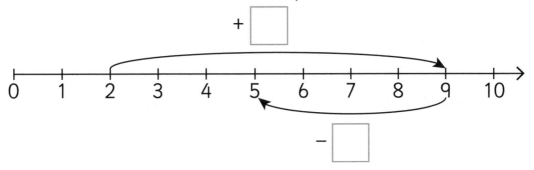

$\square \bigcirc \square \square \bigcirc \square \square = \square$

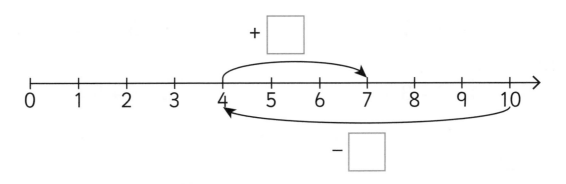

$\square \bigcirc \square \square \bigcirc \square \square = \square$

3 Write >, < or = in each $\bigcirc$.

$5 + 2 - 3 \bigcirc 8$ $8 - 4 + 2 \bigcirc 6$

$7 - 4 + 0 \bigcirc 3$ $10 + 0 - 7 \bigcirc 2$

$6 - 1 + 5 \bigcirc 4$ $9 - 4 + 2 \bigcirc 7$

4 Complete the following calculations.

$2 + 4 + 3 =$ ☐ $6 - 3 + 4 =$ ☐

$8 - 4 + 2 =$ ☐ $6 + 4 + 0 =$ ☐

$2 + 2 - 1 =$ ☐ $1 + 6 - 5 =$ ☐

$8 - 7 - 0 =$ ☐ $10 - 3 + 2 =$ ☐

$2 + 7 - 3 =$ ☐ $2 + 5 + 1 =$ ☐

$4 - 2 + 8 =$ ☐ $6 - 4 + 5 =$ ☐

Challenge and extension question

5 Think carefully and then write the answers in the boxes.

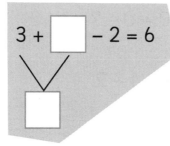

$4 + 4 -$ ☐ $= 7$

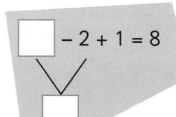

☐ $- 2 + 1 = 8$

$3 +$ ☐ $- 2 = 6$

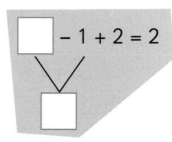

☐ $- 1 + 2 = 2$

Chapter 2 test

1 Complete the mental sums.

4 + 2 = ☐   9 – 4 = ☐   6 – 3 = ☐   5 – 5 = ☐

8 – 8 = ☐   3 + 6 = ☐   7 – 4 = ☐   6 + 4 = ☐

1 + 9 = ☐   9 – 7 = ☐   2 + 5 = ☐   10 – 5 = ☐

2 Draw lines to join the friends.

3 + 4

2 + 2

10 – 1

9 – 0

1 + 6

8 – 4

3 Look at the number lines and complete the calculations.

(a)

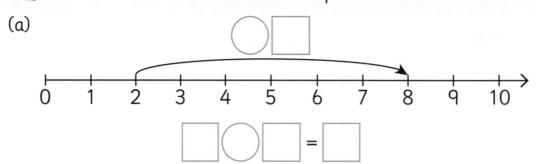

(b)

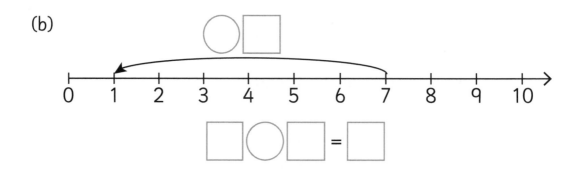

(c)

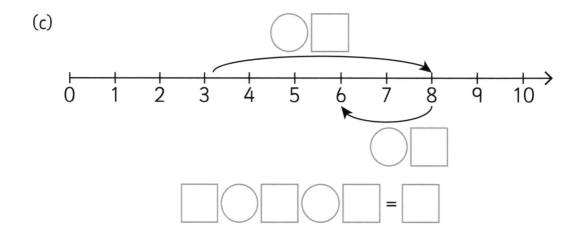

4 Draw a line to join each ball to a box.

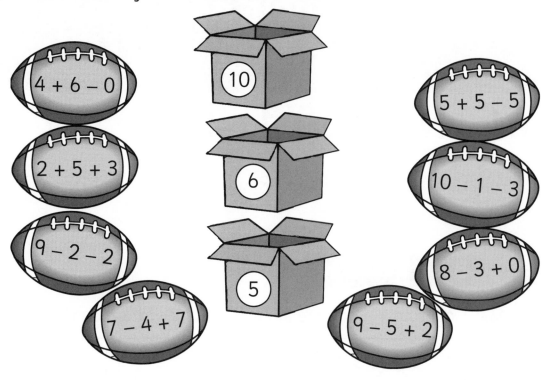

5 Look at each picture and complete the number sentence.

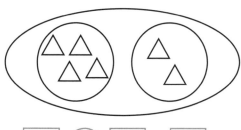

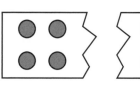
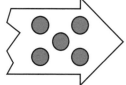

☐ ◯ ☐ = ☐   ☐ ◯ ☐ = ☐

How many leaves are
there altogether?

☐ + ☐ + ☐ = ☐

How many pieces of fruit
are left?

☐ − ☐ − ☐ = ☐

Chapter 3   Numbers up to 20 and their addition and subtraction

3.1   Numbers 11–20

 Learning objective   Read and write numbers to 20

 Basic questions

1 Count how many are in each set and write the number in the box.

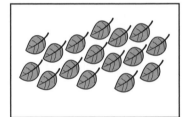

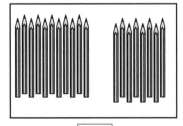

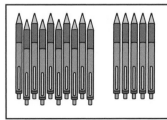

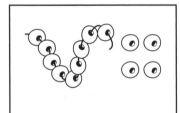

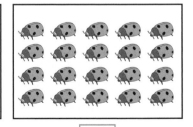

2 Fill in with numbers.

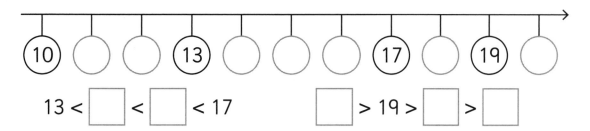

⑩ ◯ ◯ ⑬ ◯ ◯ ◯ ⑰ ◯ ⑲ ◯

13 < ☐ < ☐ < 17   ☐ > 19 > ☐ > ☐

3 Find patterns and fill in with numbers.

| 16 | 14 | | | | 6 |

| 20 | 15 | | | 0 |

⑲ ⑰ ◯ ◯ ◯ ⑨ ◯ ◯ ◯ ◯

Challenge and extension question

4 (a) The kangaroo hops on odd numbers (11, 13, ...). Help the kangaroo draw a green line to link the numbers from the least to the greatest.

The frog hops on even numbers (10, 12, ...). Help the frog draw a red line to link the numbers from the greatest to the least.

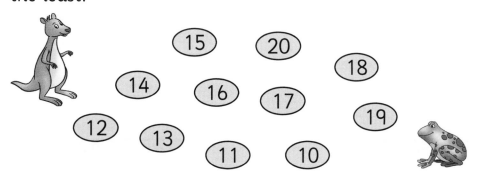

(b) Write the numbers from (a).

Odd numbers: ☐

Even numbers: ☐

3.2   Tens and ones

Learning objective Partition numbers into tens and ones

Basic questions

1 Count and draw a line to match.

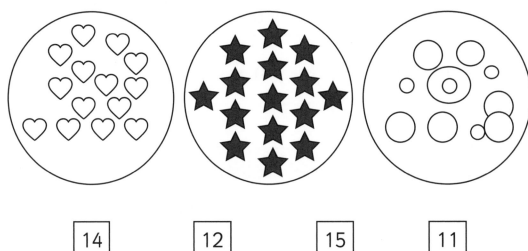

| 14 | 12 | 15 | 11 |

2 Split numbers into tens and ones. One has been done for you.

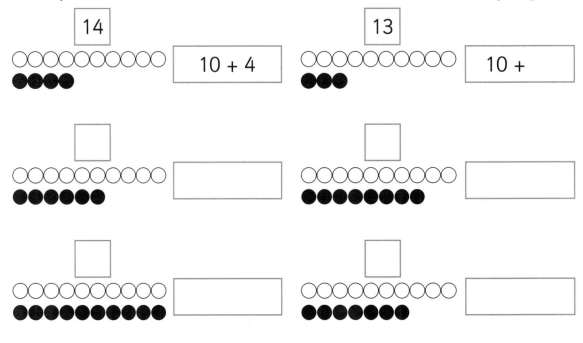

14   10 + 4   13   10 +

3 Look at each picture and write the number sentence.

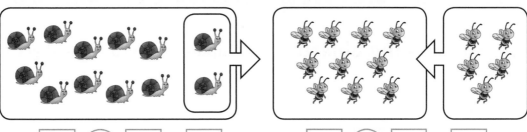

☐ ◯ ☐ = ☐   ☐ ◯ ☐ = ☐

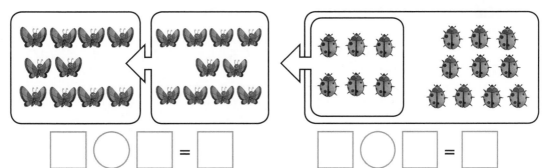

☐ ◯ ☐ = ☐   ☐ ◯ ☐ = ☐

4 Write the number in each box.

10 + 2 = ☐   10 + 5 = ☐   10 + 9 = ☐

2 + 10 = ☐   5 + 10 = ☐   9 + 10 = ☐

18 = 10 + ☐   16 = 10 + ☐   13 = 10 + ☐

20 = ☐ + 10   19 = ☐ + 9   15 = 5 + ☐

Challenge and extension question

5 Write the missing numbers in the boxes.

10 + ☐ = 14   10 + ☐ = 17   ☐ + 1 = 11

☐ + 8 = 18   6 + ☐ = 16   ☐ + 10 = 20

12 − ☐ = 10   16 − ☐ = 10

3.3 Ordering numbers up to 20

 Learning objective Compare and order numbers to 20

 Basic questions

1 Write the numbers in the circles.

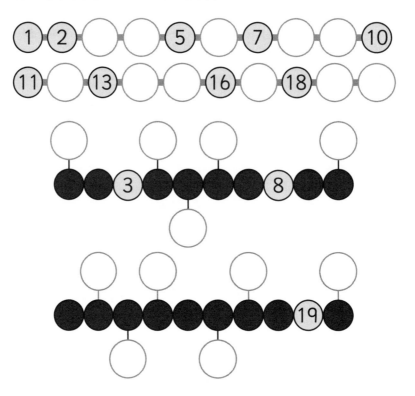

2 Count and then colour to finish the pattern on each string of beads.

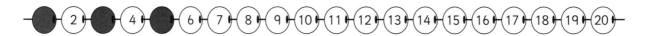

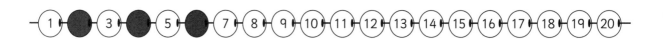

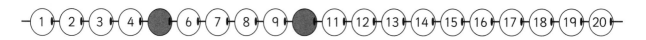

3 Write the numbers in the boxes.

| | 11 | | | 14 | | | | 18 | | |

(a) The two numbers before and after 15 are ☐ and ☐.

(b) The number after 18 is ☐ and the number before 11 is ☐.

(c) The number between 11 and 13 is ☐.

(d) The numbers greater than 12 but less than 18 are ☐.

(e) The numbers less than 20 but greater than 15 are ☐.

4 Swap the positions of two numbers in each set so the new order of the numbers forms a pattern. One has been done for you.

(a) 1, 3, 9, 7, 5   New: <u>1, 3, 5, 7, 9</u>

(b) 20, 19, 18, 17, 15, 16   New: _____

(c) 3, 6, 5, 4, 7   New: _____

(d) 4, 12, 8, 16, 20   New: _____

5 Find out and then circle the number in each set below so the remaining numbers form a pattern. One has been done for you.

(a) 3, 5, 7, ⑧, 9   Remaining: <u>3, 5, 7, 9</u>

(b) 2, 4, 6, 8, 9   Remaining: _____

(c) 18, 17, 15, 13, 11   Remaining: _____

(d) 6, 9, 12, 13, 15, 18   Remaining: _____

Challenge and extension question

6 (a) A number is greater than 12 but less than 15.

The number could be [　　　　].

(b) A number is less than 20 but greater than 16.

The number could be [　　　　].

(c) Add some numbers before and after 9 to make a number pattern.

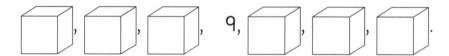

3.4 Addition and subtraction (I)

Learning objective Add and subtract using number bonds to 20

Basic questions

1 Look at the number lines and pictures and then write the number sentences.

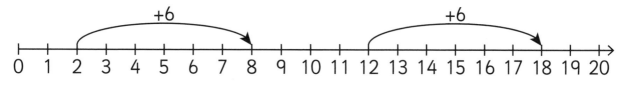

⬜ ◯ ⬜ = ⬜   ⬜ ◯ ⬜ = ⬜

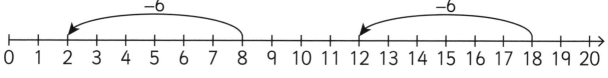

⬜ ◯ ⬜ = ⬜   ⬜ ◯ ⬜ = ⬜

⬜ ◯ ⬜ = ⬜   ⬜ ◯ ⬜ = ⬜

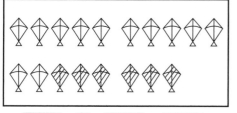

⬜ ◯ ⬜ = ⬜   ⬜ ◯ ⬜ = ⬜

2 Complete the calculations.

$6 + 2 =$ ☐ $3 + 5 =$ ☐ $4 + 4 =$ ☐ $5 + 2 =$ ☐

$16 + 2 =$ ☐ $13 + 5 =$ ☐ $14 + 4 =$ ☐ $15 + 2 =$ ☐

$1 + 5 =$ ☐ $6 + 3 =$ ☐ $3 + 3 =$ ☐ $2 + 4 =$ ☐

$11 + 5 =$ ☐ $16 + 3 =$ ☐ $13 + 3 =$ ☐ $12 + 4 =$ ☐

3 Complete the calculations.

$6 - 2 =$ ☐ $7 - 5 =$ ☐ $4 - 3 =$ ☐ $5 - 2 =$ ☐

$16 - 2 =$ ☐ $17 - 5 =$ ☐ $14 - 3 =$ ☐ $15 - 2 =$ ☐

$9 - 5 =$ ☐ $6 - 4 =$ ☐ $8 - 3 =$ ☐ $9 - 4 =$ ☐

$19 - 5 =$ ☐ $16 - 4 =$ ☐ $18 - 3 =$ ☐ $19 - 4 =$ ☐

4 Think carefully and then fill in the boxes.

$3 + 5 =$ ☐ $5 + 3 =$ ☐ $2 + 7 =$ ☐

$13 + 5 =$ ☐ ☐ $+$ ☐ $=$ ☐ ☐ $+$ ☐ $=$ ☐

$4 - 2 =$ ☐ $6 - 5 =$ ☐ $7 - 2 =$ ☐

$14 - 2 =$ ☐ ☐ $-$ ☐ $=$ ☐ ☐ $-$ ☐ $=$ ☐

Challenge and extension question

5 Write >, < or = in each ◯.

15 ◯ 2 + 13   16 – 2 ◯ 16   15 + 2 ◯ 15 – 2

13 ◯ 12 + 4   17 – 5 ◯ 10   19 – 5 ◯ 11 + 3

14 ◯ 19 – 4   13 + 7 ◯ 20   17 – 4 ◯ 17 – 3

16 ◯ 20 – 10   14 + 6 ◯ 18   2 + 16 ◯ 12 + 6

3.5 Addition and subtraction (II) (1)

Learning objective Add numbers to 20, crossing the ten and partitioning

Basic questions

1 Colour the shapes and work out the addition calculations. One has been done for you.

9 + 3 = 12

8 + 6 = ☐

7 + 5 = ☐

6 + 6 = ☐

2 Use each number line to work out the addition calculation.

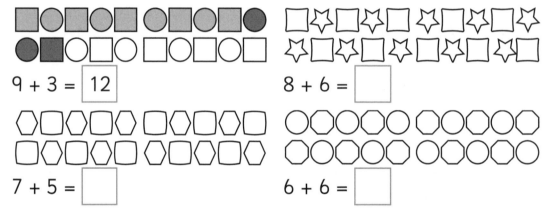

7 + 8 = ☐

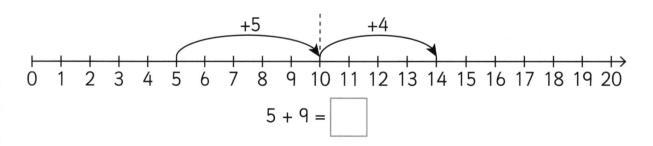

5 + 9 = ☐

3 Write the missing numbers in the boxes.

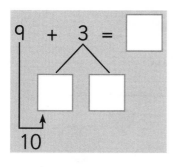

9 + 3 = ☐

☐ ☐

10

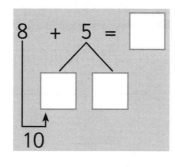

8 + 5 = ☐

☐ ☐

10

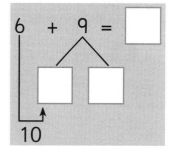

6 + 9 = ☐

☐ ☐

10

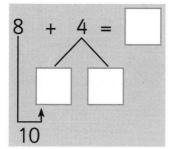

8 + 4 = ☐

☐ ☐

10

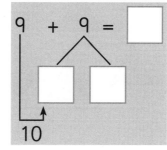

9 + 9 = ☐

☐ ☐

10

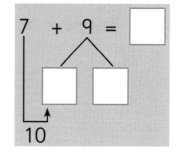

7 + 9 = ☐

☐ ☐

10

4 Find the missing numbers.

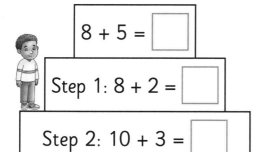

8 + 5 = ☐

Step 1: 8 + 2 = ☐

Step 2: 10 + 3 = ☐

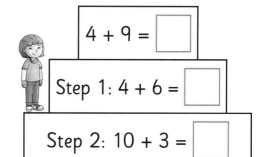

4 + 9 = ☐

Step 1: 4 + 6 = ☐

Step 2: 10 + 3 = ☐

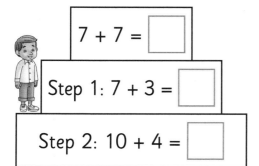

7 + 7 = ☐

Step 1: 7 + 3 = ☐

Step 2: 10 + 4 = ☐

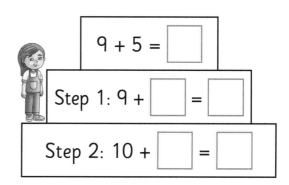

9 + 5 = ☐

Step 1: 9 + ☐ = ☐

Step 2: 10 + ☐ = ☐

Challenge and extension question

5 Use the picture to write addition sentences.

By size: _____

_____

By shape: _____

_____

3.6 Addition and subtraction (II) (2)

Learning objective Add using number bonds to 20 and the commutative law

Basic questions

1 Look at the pictures and write the number sentences.

$6 + \boxed{} = \boxed{}$

$9 + \boxed{} = \boxed{}$

$8 + \boxed{} = \boxed{}$

$\boxed{} + \boxed{} = \boxed{}$

$\boxed{} + \boxed{} = \boxed{}$

$\boxed{} + \boxed{} = \boxed{}$

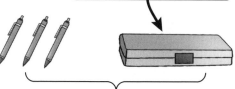

$\boxed{} + \boxed{} = \boxed{}$

$\boxed{} + \boxed{} = \boxed{}$

7 books are inside.

How many books altogether?

$\boxed{} \bigcirc \boxed{} = \boxed{}$

9 pens are inside.

How many pens in total?

$\boxed{} \bigcirc \boxed{} = \boxed{}$

2 Complete the table.

addend	8	9	5
addend	5	4	7
sum			

addend	9	3	6
addend	8	8	5
sum			

3 Work out the sums.

9 + 2 = ☐   8 + 3 = ☐   7 + 5 = ☐   4 + 9 = ☐

9 + 3 = ☐   8 + 4 = ☐   7 + 6 = ☐   3 + 8 = ☐

9 + 4 = ☐   8 + 5 = ☐   7 + 7 = ☐   3 + 9 = ☐

9 + 5 = ☐   8 + 6 = ☐   7 + 8 = ☐   2 + 9 = ☐

9 + 6 = ☐   8 + 7 = ☐   7 + 9 = ☐   9 + 7 = ☐

8 + 8 = ☐   6 + 5 = ☐   9 + 8 = ☐   8 + 9 = ☐

6 + 6 = ☐   9 + 9 = ☐   7 + 4 = ☐   6 + 7 = ☐

4 Write >, < or = in each ◯.

7 + 8 ◯ 13

4 + 9 ◯ 17

20 ◯ 13 + 7

16 ◯ 7 + 5

5 + 9 ◯ 13

8 + 7 ◯ 15

14 ◯ 8 + 4

19 ◯ 9 + 9

6 + 8 ◯ 6 + 9

5 + 8 ◯ 9 + 4

7 + 7 ◯ 10 + 4

Challenge and extension question

5 Write the missing numbers.

☐ − 3 = 7

☐ − 6 = 9

☐ − 9 = 5

☐ − 8 = 4

☐ − 7 = 5

☐ − 9 = 8

☐ − 6 = 6

☐ − 7 = 9

3.7 Addition and subtraction (II) (3)

Learning objective Subtract within 20, crossing the ten and partitioning

Basic questions

1 Cross out the sweets and then complete the subtractions.

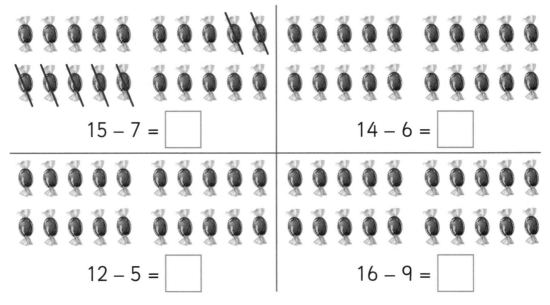

$15 - 7 = \boxed{}$

$14 - 6 = \boxed{}$

$12 - 5 = \boxed{}$

$16 - 9 = \boxed{}$

2 Use the number lines to complete the subtraction calculations.

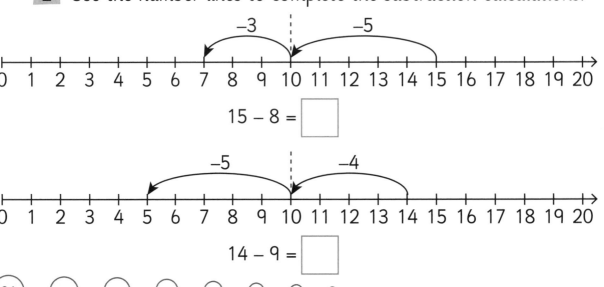

$15 - 8 = \boxed{}$

$14 - 9 = \boxed{}$

3 Write the missing numbers in the boxes.

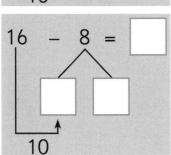

12 – 3 = ☐
☐ ☐
10

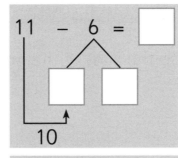

11 – 6 = ☐
☐ ☐
10

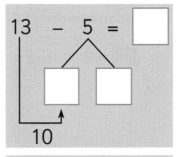

13 – 5 = ☐
☐ ☐
10

16 – 8 = ☐
☐ ☐
10

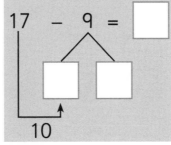

17 – 9 = ☐
☐ ☐
10

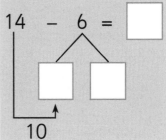

14 – 6 = ☐
☐ ☐
10

4 Fill in the missing numbers.

14 – 5 = ☐
Step 1: 14 – 4 = ☐
Step 2: 10 – 1 = ☐

12 – 9 = ☐
Step 1: 12 – 2 = ☐
Step 2: 10 – 7 = ☐

15 – 7 = ☐
Step 1: 15 – 5 = ☐
Step 2: 10 – 2 = ☐

11 – 9 = ☐
Step 1: 11 – 1 = ☐
Step 2: 10 – ☐ = ☐

Challenge and extension question

5 Fill in the ( ) so the sum of the numbers on each line is 20.

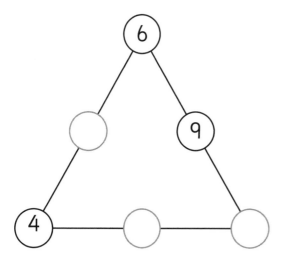

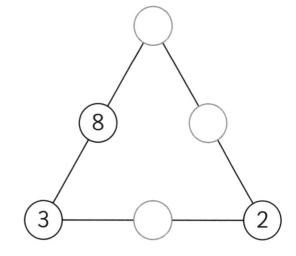

3.8 Addition and subtraction (II) (4)

Learning objective Subtract numbers within 20

Basic questions

1 Look at the pictures and write the number sentences.

13 − 5 = ☐

15 − ☐ = ☐

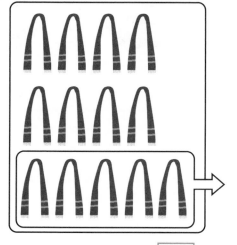
☐ − ☐ = ☐

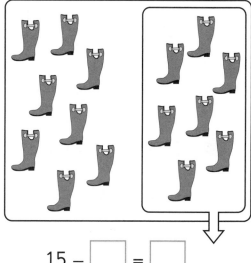
☐ − ☐ = ☐

How many buttons are inside?

How many pens are inside?

There are 13 buttons altogether.

☐ ◯ ☐ = ☐

There are 12 pens in total.

☐ ◯ ☐ = ☐

There are 16 pieces
of fruit altogether.

9 apples   How many?

☐ ◯ ☐ = ☐

There are 20 cars in total.

9 cars   How many?

☐ ◯ ☐ = ☐

2 Complete the table.

minuend	11	13	15
subtrahend	5	4	7
difference			

minuend	16	12	14
subtrahend	8	9	6
difference			

3 Complete the subtraction calculations.

11 – 2 = ☐ 11 – 8 = ☐ 12 – 5 = ☐ 13 – 8 = ☐

12 – 3 = ☐ 12 – 4 = ☐ 14 – 6 = ☐ 16 – 9 = ☐

13 – 4 = ☐ 12 – 6 = ☐ 15 – 7 = ☐ 13 – 6 = ☐

14 – 5 = ☐ 14 – 7 = ☐ 12 – 8 = ☐ 12 – 7 = ☐

4 Write >, < or = in each ◯.

13 – 8 ◯ 7   14 – 9 ◯ 7   6 ◯ 13 – 7   9 ◯ 14 – 6

14 – 5 ◯ 8   16 – 9 ◯ 5   10 ◯ 18 – 9   7 ◯ 11 – 9

Challenge and extension question

5 Look at the numbers first and then fill in the boxes.

8 – 1 = 7   ☐ – ☐ = 7   ☐ – ☐ = 7

9 – 2 = 7   ☐ – ☐ = 7   ☐ – ☐ = 7

10 – 3 = 7   ☐ – ☐ = 7   ☐ – ☐ = 7

☐ – ☐ = 7   ☐ – ☐ = 7   ☐ – ☐ = 7

3.9 Addition and subtraction (II) (5)

Learning objective Add and subtract numbers to 20

Basic questions

1 Look at the pictures and write the number sentences.

 ⑦ ⑤

How many rabbits are there in total?

☐ ◯ ☐ = ☐

How many are in the basket?

There are 11 carrots in total.

☐ ◯ ☐ = ☐

How many are inside?

There are 13 flowers in total.

☐ ◯ ☐ = ☐

☐ ◯ ☐ = ☐

 9 people are on the bus.

How many people are there in total?

☐ ◯ ☐ = ☐

4 apples were eaten.

How many people were there in the first place?

☐ ◯ ☐ = ☐

How many are left?

There were 12 rabbits at first.

☐ ◯ ☐ = ☐   ☐ ◯ ☐ = ☐

2 Complete the following calculations.

12 − 0 = ☐   5 + 7 = ☐   15 − 8 = ☐   13 − 5 = ☐

9 + 6 = ☐   11 − 3 = ☐   9 + 0 = ☐   12 − 4 = ☐

7 + 6 = ☐   14 − 7 = ☐   13 − 8 = ☐   19 + 1 = ☐

3 Work out the calculations on the boxes and then draw a line to match each one to the correct house.

Result greater than 10

Result less than 10

5 + 7   15 − 6   7 + 13   9 + 2

16 − 5   8 + 4   13 − 8   18 − 14

4 Write the numbers in the boxes.

13 $\xrightarrow{-6}$ ☐   8 $\xrightarrow{+4}$ ☐   14 $\xrightarrow{-5}$ ☐

17 $\xrightarrow{-4}$ ☐   6 $\xrightarrow{+9}$ ☐   9 $\xrightarrow{+8}$ ☐

☐ $\xrightarrow{-7}$ 6   12 $\xrightarrow{-}$ 4   11 $\xrightarrow{-}$ 6

 Challenge and extension question

5 Choose six of these numbers and then use them to fill in the boxes so that the equations are correct.

③ ④ ⑤ ⑥ ⑦ ⑧ ⑨

☐ + ☐ = ☐ + ☐ = ☐ + ☐

3.10　Let's talk and calculate (III)

Learning objective　Interpret word problems using addition and subtraction facts to 20

Basic questions

1　Read each problem and write the addition or subtraction sentences.

There were 16 🍎 on a plate.
7 🍎 were eaten.
How many 🍎 were left?

☐ ◯ ☐ = ☐

There were 12 🍐 on a table.
9 🍐 were taken away.
How many 🍐 were left on the table?

☐ ◯ ☐ = ☐

6 🐦 were in a tree.
Another 5 🐦 arrived.
How many 🐦 are in the tree now?

☐ ◯ ☐ = ☐

9 🚗 were in a car park.
Another 7 🚗 drove into the car park.
How many 🚗 are in the car park now?

☐ ◯ ☐ = ☐

There are 6 .

There are 14 .

How many more are needed so there are as many as ?

☐ ◯ ☐ = ☐

There are 7 .

There are 5 .

How many and are there in total?

☐ ◯ ☐ = ☐

2 Read each problem and then write the number sentence.

(a) There were 6 lambs in the field. Another 7 lambs joined them. How many lambs are there altogether?

Number sentence: _____

(b) Theo has 6 yellow pencils. He has as many blue pencils as yellow pencils. How many pencils does he have altogether?

Number sentence: _____

(c) Orla's sister gave her 12 cherries. She ate 3 of them. How many cherries were left?

Number sentence: _____

(d) 14 children were playing in the park. 8 of them went home. How many children stayed in the park?

Number sentence: _____

3 Read each problem and then write the number sentence.

(a) Jake needs to make 12 cakes. He has made 8. How many more does he need to make?

Number sentence: _____

(b) There are 9 grapes left after Tom has eaten 6 grapes. How many grapes were there at first?

Number sentence: _____

Challenge and extension question

4 Add a condition to complete the story. Then write the number sentences and find the answers.

(a) There are 8 red pens.

_____.

How many red and green pens are there altogether?

_____

(b) _____.

9 cars drove away. How many cars are left?

_____

3.11 Adding on and taking away

Learning objective Use the inverse relationship between addition and subtraction with numbers to 20

Basic questions

1 Look at the pictures and fill in the boxes.

```
        5            10           15           20
●●●●●●  ● ○○○○○  ○○○○○○  ○○○○○
        ⊘⊘⊘⊘⊘  ⊘⊘⊘⊘⊘
```

6 + 9 = ☐   15 − 9 = ☐

```
        5            10           15           20
●●●●●●  ●●●●●○  ○○○○○○  ○○○○○
        ⊘  ⊘⊘⊘⊘⊘  ⊘
```

9 + ☐ = ☐   ☐ − ☐ = ☐

```
        5            10           15           20
●●●●●●  ●●●○○  ○○○○○○  ○○○○○
        ⊘⊘  ⊘⊘⊘
```

☐ + ☐ = ☐   ☐ − ☐ = ☐

2 Use the pictures to write addition and subtraction sentences.

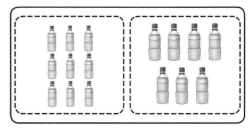

☐ + ☐ = ☐

☐ − ☐ = ☐

☐ + ☐ = ☐

☐ − ☐ = ☐

3 Find the missing number for each ☐ and
write + or − in each ◯.

15 ⟷ ☐
    −3

17 ⟷ ☐
    − ☐
    +2

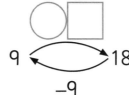
9 ⟷ 18
   −9

14 ⟷ ☐
    +5
◯ ☐

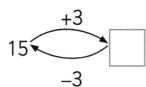
13 ⟷ 14
◯ ☐

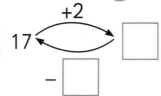
7 ⟷ 15
◯ ☐

8 ⟷ ☐
   −4
   +4

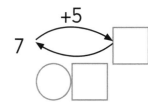
7 ⟷ ☐
   +5
◯ ☐

4 Think carefully and then fill in the boxes.

16 + 2 = ☐	☐ + 4 = ☐	18 − 9 = ☐
18 − 2 = ☐	14 − 4 = ☐	☐ + 9 = ☐
7 + 4 = ☐	16 − 8 = ☐	13 − 4 = ☐
11 − ☐ = ☐	☐ + 8 = ☐	☐ + ☐ = ☐

Challenge and extension question

5 Read each number story and then write the question. Then write the number sentence.

(a) There were 18 children on the train. 4 of them got off at

a station. _____?

Number sentence: _____

□ ○ □ = □

(b) Min bought 15 books, and then she bought another

4 books. _____?

Number sentence: _____

□ ○ □ = □

(c) There were 14 cars in the car park at first. 8 of them then drove away. _____?

Number sentence: _____

☐ ◯ ☐ = ☐

(d) Amber had 13 strawberries. She gave 8 to Jacob.

_____?

Number sentence: _____

☐ ◯ ☐ = ☐

3.12 Number walls

Learning objective Use number bonds to 20

Basic questions

1 Fill in the number walls. One has been done for you.

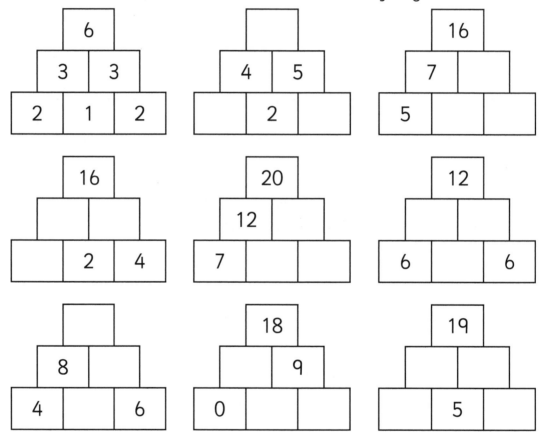

2 Fill in the number walls.

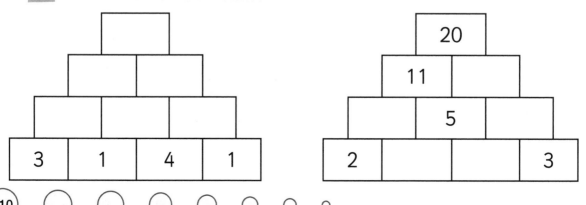

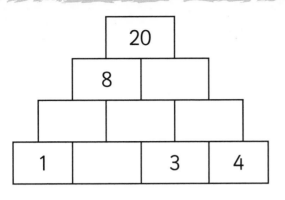

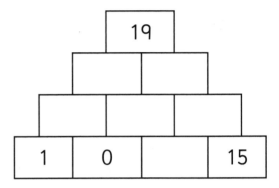

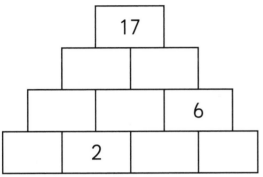

Challenge and extension question

3 Build your own number walls.

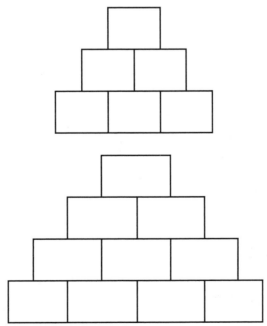

Chapter 3 test

1 Fill in the boxes.

(a) Write the missing number in each box on the number line.

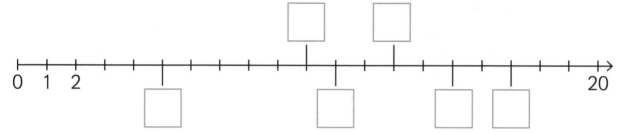

(b) Fill in the missing numbers to complete the number patterns.

2 , 4 , 6 , ☐ , 10 , ☐ , ☐ .

19 , 17 , 15 , ☐ , 11 , ☐ , ☐ .

(c) Write the number on each sail.

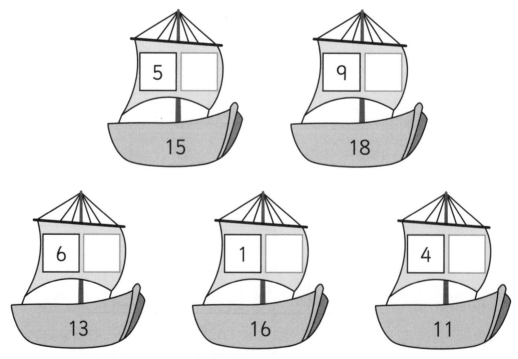

2 Write the missing numbers.

(a) Write in the boxes.

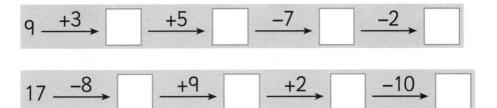

9 $\xrightarrow{+3}$ ☐ $\xrightarrow{+5}$ ☐ $\xrightarrow{-7}$ ☐ $\xrightarrow{-2}$ ☐

17 $\xrightarrow{-8}$ ☐ $\xrightarrow{+9}$ ☐ $\xrightarrow{+2}$ ☐ $\xrightarrow{-10}$ ☐

(b) Fill in the number walls.

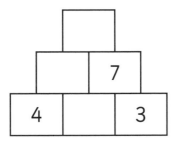

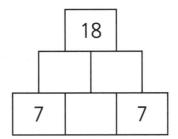

(c) Draw a line to match the birds to their homes.

15 – 7

17 – 9

6 + 4

8

10

16

8 + 8

13 – 5

14 – 4

10 + 6

3 Look at the pictures and write the number sentences.

7 + ☐ = ☐

13 − ☐ = ☐

☐ + ☐ = ☐

☐ − ☐ = ☐

How many apples are inside?

There are 16 apples in total.

☐ ◯ ☐ = ☐

There are 11 books inside

How many books in total?

☐ ◯ ☐ = ☐

4 Fill in the boxes.

If 🎈 + 🪁 = 17 and 🪁 + 🪁 = 18, then

🎈 = ☐ , 🪁 = ☐ .

Notes

Notes